CHASING SCIENCE

Books by Frederik Pohl

THE HEECHEE SAGA
Gateway
Beyond the Blue Event Horizon
Heechee Rendezvous
The Annals of the Heechee
The Gateway Trip

THE ESCHATON SEQUENCE
**The Other End of Time*
**The Siege of Eternity*
**The Far Shore of Time*

The Age of the Pussyfoot
Drunkard's Walk
Black Star Rising
The Cool War
Homegoing
Mining the Oort
Narabedla Ltd.
Pohlstars
Starburst
The World at the End of Time
Jem
Midas World
Merchant's War
The Coming of the Quantum Cats
Man Plus
Chernobyl
The Day the Martians Came
Stopping at Slowyear
**The Voices of Heaven*
**O Pioneer!*

WITH JACK WILLIAMSON:
The Starchild Trilogy
Undersea City
Undersea Quest
Undersea Fleet
Wall Around a Star
The Farthest Star
**Land's End*
The Singers of Time

WITH ISAAC ASIMOV:
Our Angry Earth

WITH LESTER DEL REY:
Preferred Risk

WITH C. M. KORNBLUTH:
The Space Merchants
Gladiator-at-Law
Search the Sky

The Best of Frederik Pohl
(edited by Lester Del Rey)
The Best of C. M. Kornbluth
(editor)

NONFICTION:
The Way the Future Was
**Chasing Science*

*denotes a Tor book

CHASING SCIENCE

SCIENCE AS SPECTATOR SPORT

Frederik Pohl

A Tom Doherty Associates Book
New York

CHASING SCIENCE

This book is printed on acid-free paper.

Book design by Jane Adele Regina

Edited by James Frenkel

A Tor Book
Published by Tom Doherty Associates, LLC
175 Fifth Avenue
New York, NY 10010

www.tor.com

Library of Congress Cataloging-in-Publication Data

Pohl, Frederik.
Chasing science : science as spectator sport / Frederik Pohl—1st ed.
p. cm.
ISBN 0-312-86711-5
1. Science—Popular works. I. Title.

Q162.P5914 2000
500—dc21 00-057768

First Edition: December 2000

Printed in the United States of America

10 9 8 7 6 5 4 3 2 1

For Lester del Rey,
Jack Williamson, Betty Anne Hull,
and all the others who shared the chase.

Contents

INTRODUCTION: A LOVE AFFAIR

How I Got Switched on to Science for the Fun of It

Thirty-odd years ago the science fiction club at the Massachusetts Institute of Technology invited me to come up and deliver a lecture for them. I did. Three or four hundred techies turned up to hear me, and the talk went pretty well. The questions and discussions were good; and then, when I was preparing to leave, a young man came up and introduced himself. He was a graduate student in computer studies, he said, and as long as I was there, would I like to see what was going on in MIT's computer lab? I said sure, and he escorted me across the campus to the lab.

Remember, this was in the early 1960s. Computers were still pretty new, and by today's standards pretty feeble, too; my current laptop has far more computing power than anything MIT owned in those days. I didn't know an awful lot about computers at the time, either. I was aware that they crunched numbers more efficiently than any human being. I even knew that, in order to do so, they converted all the arithmetic into binary notation—that is, counting to the base 2 instead of to the base 10 that we humans use—and I was aware that in order to do so, each computer contained little transistors that could either be on or off, thus representing either a 1 or a 0 in binary.

That, however, was about *all* I knew until the people at the computer lab began to open my eyes.

The first thing they showed me was a game. I hadn't known you could play games on a computer, especially not games of kinds that you couldn't play anywhere else. To run this particular game took a state-of-the-art PDP computer, a gray metal brick about the size of a suitcase, with a screen about the size of a dinner plate, and the game was called Space War. The screen of the computer displayed a background of points of light—representing stars—and moving around the starfield were two squat little rockets. The stars stayed still; the rockets were movable. Each of them had a little hand control that could make it go forward or turn, with an extra little button for launching a space torpedo at the other one; when a torpedo hit, it exploded in a flare of light, meaning victory for the ship that launched the torpedo.

That doesn't sound like much compared with some of the things in any present-day arcade, but in the early 1960s it was fascinating. (Still is, as a matter of fact. That is why playing Space War is no longer permitted in the MIT computer labs, since it keeps the graduate students from doing what they're supposed to be doing there.) I played about a dozen rounds and was shot down in flames every time. Well, what could you expect? I was a rookie, while every one of the students had had many hours of practice before I came along.

Then they showed me some other things. What, they asked, did I think the universe would be like if the gravitational attraction between two bodies worked at right angles to the line between them? Since I didn't know, they showed me. A graduate student named Marvin Minsky had instructed a computer to display how three particles would interact under those conditions. What came out of that they called the "Minskytron," and what it looked like was a rapidly whirling kaleidoscope. And that was only the beginning. Then, and in later visits (for I came back to that lab as often as I could), I watched their first attempts to build a robot eye that could recognize what it saw and a robot hand that could pick objects up and move them about; factored some large numbers into primes; even was allowed to play with MIT's large central computer when it was fitted with workstations all over the campus; learned about the attempts by Minsky (who had become a good friend) and others to teach computers to pick out the moral from certain children's stories; and generally had a fine time there. I still do, in fact, because only a couple of months ago I was invited to take part in a symposium there on wearable computers. (Wearable computers! So far we have come from the PDP.)

I couldn't stay away. On the train home after that first visit, I realized I had fallen hopelessly in love.

I'VE STAYED IN love, too. Over the years many things have changed in my life, but that isn't one of them. Science is still my favorite recreation, all kinds of science, whether I find it in a laboratory, an observatory, on the slope of a volcano, or in my own backyard.

I should explain that I am not in any sense a scientist myself.

Apart from the science courses I took at Brooklyn Technical High School as a teenager and what I learned from the air force's training to be a weatherman in World War II, I have had no formal scientific education at all.

What I am is a *fan* of science. My relation to science is the same as my relation to the New York Mets. I don't hope to make the team, I just like to watch them play—the difference being that, for me, science is even better as a spectator sport than baseball ever was.

One of the great fringe benefits of being a professional writer is that it gives me the chance to do a good deal of traveling. I've used that opportunity to chase science all over the world—in every state of the U.S.A., most Canadian provinces, six of the seven continents (I've never been to Antarctica and never will, because tourism is terribly destructive of its fragile ecology), and somewhere around fifty foreign countries. I can't think of anything that would make me abandon that loving pursuit short of total bodily paralysis or death.

What's more, I, can't really understand why there are any human beings alive in the world today who don't share my infatuation with the subject.

I do know what people *say* to excuse the fact that they shut their eyes to science. One frequent complaint is that science is hard to understand, which is at least sometimes true when you explore its furthest reaches. I would not deny that it is not at all easy to comprehend, for instance, some of the spookier parts of relativity, biochemistry, or quantum physics. But you don't have to pass a written examination in astronomy to feel a thrill when some new picture comes in from a spacecraft near a distant planet . . . and, anyway, there are not very many basic principles in science that are much harder to learn than the vast quantity of arcane sports lore that every ten-year-old readily commits to memory, from basketball statistics to the infield fly rule.

Science isn't just made up of big machines and complicated equations. Science is much simpler and more beautiful than that. At its root, science is really nothing more than a systematic process of looking at the world around us—all of it, including its furthest reaches into time and space—and trying to figure out what the rules are that make the whole thing tick. And, really, are there that many better things for anyone to do with his life, or hers?

I don't think so . . . as you can see by the fact that it's what I've done with so much of mine.

* * *

THIS ISN'T A scientific text, and if there are scientific errors in it, as well there may be, I apologize. What I intend by this book is something quite different. I've had a great deal of fun chasing science, over many years, and the reason I wrote this book is that I would like to share some of that pleasure with you.

Chapter 1
THE PLACES WHERE SCIENCE HAPPENS

National Laboratories and Others

Some of the best places to chase science are the places where scientists are actually doing it. Some of those places are, that is. Not all of them are equally pleasurable.

The thing is that there are two ways to do science. One way is to make experiments and observations to see what is so about the world. The other way is to think about what the experimenters and observers have found and to try to construct theories to explain it. The two modes are equally important in trying to increase our understanding of the universe.

They are not, however, equal in entertainment value. When a theoretician theorizes, he or she usually does so in front of a whiteboard or a desktop computer. The principal activity going on there is thinking. Watching it happen is just about as exciting as watching paint dry. For the real fun part of science you need to go to where the experiments are being done, and that is the laboratories.

When we think of "laboratories," the picture in our minds may look a lot like Dr. Frankenstein's old movie set, with its rising, widening Jacob's ladder of electrical sparks, blended with (in my own case) memories of my first Brooklyn Tech chemistry course, in the old factory building at the eastern end of the Manhattan Bridge. Those labs were the first real ones I had ever seen. They were instantly identifiable by their zinc sinks and strange liquids bubbling over Bunsen burners, along with the persistent faint vinegary smell of "experiments" that always had a foreknown result.

The actual laboratories of real present-day scientists are seldom like that. They are a lot more complicated, and, unless the scientist is a biochemist or perhaps a veterinary researcher, they hardly ever smell.

Probably the best labs to start out with are the country's national laboratories, owned and operated by the government of the United States and therefore, if you are a taxpaying voter, by you. This means that almost all of them are likely at the very least to run occasional guided tours. They pretty much have to, you see. They need government funding to keep going, and they prudently cultivate the goodwill of voters like you to make sure the money taps stay open.

The exceptions to this rule are the labs predominantly engaged in classified military research. Even those may permit a few closely watched visitors from time to time, so it's worth a try to place a phone call to the laboratory's public information office and inquire about the possibility of a tour . . . but don't count on being welcomed with open arms.

However, the arms of at least one national laboratory are always open and welcoming. That's Fermilab, located in Batavia, Illinois, an hour's drive southwest of Chicago.

When you feel like going to Fermilab, you don't have to plan very far ahead. You can just wander in off the street, without asking anyone's by-your-leave, and still get a good flavor of what's going on. Once you're inside the main building, Wilson Hall (easy to recognize, because it is by far the tallest building on Fermilab's 640-acre campus), you take the elevator up to the fifteenth floor of the tower. There you'll find a twenty-minute video that shows what Fermilab is all about. When that's over you just pick up a tape-playing headset and then stroll along a visitors' gallery where the tape will explain what you are seeing: namely actual sections of the piping, the magnets, and all the other parts of the great underground accelerator ring that is Fermilab's principal reason for being.

While you're up there you can look out toward the distant Chicago skyline, off toward the northeast, and when you've looked your fill you might go back down to the lobby and inquire what that evening's entertainment is going to be. Fermilab runs a regular lecture series in its auditorium. The lecturers are scientists, and the principal purpose of the series is to keep Fermilab's own people in touch with what is going on elsewhere. Still, when seats are available the public is welcome to come in and listen; so, too, with the concerts and films that run there when no lecture is scheduled.

If it turns out that there is something coming up that you'd like to see and you choose to wait around for it, you can pass the time by having a meal or a cup of coffee in the cafeteria on the ground floor of Wilson Hall's magnificent atrium. The food is edible, the ambience fine and the company great. What's more, if you are inspired with a great new idea in the middle of your meal, you can do what the particle physicists at the next table are doing and jot down your inspiration on one of the little scratch pads the management has put at each place, to spare wear and tear on their tablecloths.

What you can't see for yourself is the actual Main Ring particle accelerator, because it's buried underground. That's not all bad, though. You wouldn't really want to get too close to it while it was in operation, because the radiation it produces would kill you.

FERMILAB'S BASIC BUSINESS is what used to be known as atom smashing. However, the atom itself has been well and truly smashed long since, and its basic pieces identified: There's the proton, the positively charged particle in the core of the atom; the electrically neutral neutron, also in the core; and the negatively charged electron, much smaller than the other two and residing in shells around the core.

Those are the basic building blocks of the chemical elements (of which all matter is composed), and they're simple enough. An atom made up of one proton and one electron is hydrogen, lightest of the elements. If the atom has two protons and two electrons, it's helium, the next lightest, and so on up to uranium, heaviest of the natural elements, with ninety-two of each.

What complicates the picture is the neutron. If to your single proton you add a neutron in the core, there's still only one electron, and so, chemically, what you get is still hydrogen. But it's twice as heavy as the previous atom of hydrogen so it's called heavy hydrogen or deuterium. If you add two neutrons, it hasn't stopped being hydrogen but now it's heavier than ever; it is now called tritium, and although its chemistry hasn't changed, it is now radioactive and in fact forms an important part of the hydrogen bomb.

Protons, neutrons, and electrons are reasonably user-friendly particles. If the roll call stopped with them, physicists wouldn't have much to do. Scientists have good ways of detecting those particles, and even of counting and weighing them, so that they now know that it takes 1,841 electrons to weigh as much as a single proton, for instance.

What's more, the electrons, protons, and neutrons have another comforting trait. They stick around for a good long time. Electrons basically never die. Protons *may* die after some very long time—scientists are still uncertain about that—but for normal observational purposes they're close enough to immortal to be not worth worrying about.

This is all old stuff; even the neutron, the most recently identified of the three, was discovered while Herbert Hoover was still President of the United States.

What has changed is that it is now clear that there are many other particles around, some with lifetimes of less than a billionth of a billionth of a second, some electrically neutral and so small that they are almost impossible to detect (like the various kinds of neutrinos). Even more astonishing, it turns out that the proton itself is not just a simple lump of positive electricity. It is made up of still smaller particles called quarks, and it is the search for exotic particles like these that makes up the work of Fermilab's Main Ring.

Before you can study these particles you need to find a way of creating them in a place where they will register on detectors. The way to do that is to speed up streams of other particles—more easily handled ones, like the protons that are in every atom—as fast as you can make them go, even close to the limiting speed of light. And then you crash them into some kind of a target and see what fragments fly off.

There's nothing theoretically hard about accelerating nuclear particles. We all have particle accelerators in our homes, right in our TV sets: They are what launch electrons to strike the phosphors on the inner surface of the picture tube with enough force to make them flash into light, thus giving us the pictures we watch. But doing it on the scale necessary for studying exotic particles is a serious engineering problem. The usual technique is to dig a circular trench around your laboratory, lay a metal pipe in the trench, pump all the air out of the pipe, and fit it with rings of magnets all around its perimeter. Then you bury your trench (to protect it from the weather, but also because it will produce some kinds of radiation that you don't want your scientists exposed to) and start loading it with clusters of particles—say, with protons, for a starter. The magnet rings around the pipe go on and off in rapid succession; each set of magnets pulls the clump of particles toward it and goes off in time to turn it over to the next set. When you've accelerated your particles as fast as you can get them to go, you redirect their final circuit of the ring and smash them into a metal target; then you look to see what fragments have been produced.

None of that is easy. The engineering is expensive and difficult. It takes work to get a good enough vacuum in the pipes to make the accelerator work in the first place—even a tenuous wisp of leftover gas will scatter the particles into uselessness. It's hard to build magnets that are powerful enough, and can be switched on and off fast enough, to accelerate particle streams up to the required speeds. The only kind of magnets that will do the job operate at very cold temperatures, close to absolute zero, so you need to refrigerate them—liquid helium is the refrigerant of choice. (For that reason, one of the buildings you saw when you were on the fifteenth floor is a liquid-helium factory.) And it's hard to keep your particles in compact clusters, because they have no reason to want to stay together. What they want is to spread out and expand, and you've got to work hard to keep them focused.

Particle accelerators are classical examples of Big Science. That kind of construction costs a lot of money, and it takes up a lot of room as well.

The Main Ring accelerator at Fermilab has a diameter of well over a mile. It encloses so much otherwise wasted parkland space—a square mile of it within the Main Ring, another six thousand acres surrounding the lab itself—that the Fermilab authorities (who, like most scientists, are eager conservationists) have made thrifty use of the pasture. They have stocked it with a thousand white-tailed deer and eighty-five American buffalo, and they've planted it with the native prairie grasses and wildflowers that flourished in the area before human beings changed it all. There are lakes and streams and woods to attract hikers, birdwatchers, fishermen, and picnickers. Plenty of them use the site regularly.

Not all the wildlife at Fermilab is there by design. There are also ducks, geese, songbirds, and raccoons, and sometimes they cause trouble. Not long ago a raccoon squeezed into a junction box for the Main Ring, and a goose somehow got itself into a power line. The resulting short circuits were bad for the scientists—their experiments were shut down until repairs could be made—but, of course, a lot worse for the raccoon and the goose.

One other thing about Fermilab hospitality. From time to time the management declares a Saturday open house. Many of the working scientists show up as volunteer guides for whoever cares to visit, and if you are present you will be taken to the working labs, the place where the liquid helium is made, and the tunnels under Wilson Hall, where you will see, among other things, the ports that allow carefully monitored amounts of radiation to escape for the treatment of cancer patients.

Taken all in all, Fermilab is a nice place—due largely, in my opinion, to the fact that the man who ran the place for many years was Leon Lederman, who is not only a brilliant scientist but a particularly amiable one.

BROOKHAVEN NATIONAL LABORATORY, on Long Island, New York, isn't quite as user-friendly as Fermilab. You can't just walk in when you feel like it. However, you needn't be excluded, either. Brookhaven runs fairly frequent group tours, and if you call the public information office you may be allowed to tag along on one of them.

Brookhaven's range of research is surprisingly broad. One team produces some of the country's most advanced planning on magnetic-levitation transportation, another investigates what goes on in the brains of drunks by means of PET (short for positron emission tomography). However, like Fermilab, its particle accelerator is its main set piece. Brookhaven's is smaller, but it does similar research into subatomic particles. (I emphasize the word "similar." This is not a case of wasteful government duplication. Each research facility has its own agenda and its own special strengths.)

When I drove into the Brookhaven complex recently I realized I had been there before. By "before" I mean quite a while before, even before it became a national laboratory. Actually, I had spent a rather stressful week on those grounds during World War II. It was called Camp Upton then, and it served as the reception center for New York–area civilians as they were being turned into newly minted soldiers. I don't think I would have recognized the place if I hadn't known what it was. It had changed a lot. The army's wood-frame barracks and mess halls were gone, replaced by structures that looked more like the campus of a community college than anything else.

I did recognize the surroundings, because Brookhaven/Camp Upton is set in prettily rolling country—which is not surprising when you bear in mind that, after all, all of Long Island is only the hilly rubbish heap the great glaciers pushed ahead of them as they slid southward in the last ice age. There's another particularly attractive thing about touring Brookhaven. Its accelerator, being smaller than Fermilab's, doesn't present the same sort of danger from seeping radiation. Therefore much more of its workings are visible to the visitor.

And Brookhaven has one unique trait. Every other accelerator ring in the world runs clockwise. Brookhaven's runs the other way.

There isn't any particular reason for this. It makes little difference to the particles which way they spin, and the Brookhaven accelerator's orientation was simply an arbitrary decision of the architects. However, the Brookhaven crew got a little tired of hearing visiting physicists from Fermilab or Europe's CERN sneer at their backward ring. They couldn't reverse the ring. What they could do, and did, was reengineer their control-room clock.

Now their laboratory clock runs counterclockwise, and when visitors comment on the handedness of the Brookhaven particle accelerator, its scientists are quick to point out that theirs is the only ring in the world that does run clockwise—by their clock.

NOT ALL NATIONAL laboratories concern themselves with particle acceleration. One with a different mandate, and a completely different history, is the Argonne National Laboratory in Illinois, a little farther south and west of Chicago than Fermilab.

Argonne's birth was almost accidental. In the early winter of 1942, Enrico Fermi, Leo Szilard, and a number of other leading experts in the physics of the atomic nucleus were busy trying to find out if an atomic bomb was possible. The first step was to see if a self-sustaining atomic reaction could be achieved at all. So, in an abandoned squash court under the grandstand for the University of Chicago's football stadium, Stagg Field, they laboriously constructed the first such reactor. Since their initial design involved piling many carbon bricks around the uranium fuel, they called it a pile; and, on December 2, 1942, it succeeded in releasing some power. Not much. Only about 1 watt of power, in fact, but that was enough to show that the principle was sound, and shortly thereafter the mammoth Manhattan Project was begun to take over the task of actually building a bomb.

Once the Stagg Field experiment was successful, the authorities at the University of Chicago began to get nervous about having a working nuclear reactor right in the middle of their campus. So the pile was carefully disassembled, and it, with all its associated systems and instrumentation, was taken to a wooded area far enough from the city to cause no fears. That area had been named the Argonne Forest—memorializing one of the great American battles of World War I—and it was located in the Palos Hills Forest Preserve. That was the beginning of the Argonne National Laboratory.

Half a century later, the laboratory still does some research into nuclear power, but that is only one program of the scores which are under way there. Besides designing the first prototype nuclear-power submarine engine, Argonne's scientists have designed improved

Braille machines for the blind, planned a diet to reduce the effects of jet lag, and done significant work in so-called high-temperature superconductors. ("High" in this context refers to superconductors that work at a temperature as much as thirty or forty degrees above absolute zero.) And their newest and biggest current program is the Advanced Photon Source, the world's most powerful X-ray generator of its kind. With X rays billions of times more powerful than the ones used in medicine, it is capable of studying the arrangement of atoms in molecules, with as many as seventy experiments going on at once.

I have a couple of favorite personal memories of Argonne. For one thing, although I had been hearing about electron microscopes for years, it was there that I actually saw one for the first time—I had not realized that the things could be two or three stories tall. Even more interesting was what the Argonne people call their Cave.

The Cave is a cubicle about the size of an apartment kitchenette, and it is the heart of Argonne's virtual-reality program.

To get the visual illusion that you are actually in some strange place while in fact you are simply standing in the Cave takes some powerful computer time, but you won't see any of that. It's all behind the scenes. The visible part of Argonne's virtual-reality system is nothing but a pair of goggles. They're pretty high-tech glasses, though. As the computer feeds them images of whatever it is you choose to see, each glass blinks alternately on and off, so rapidly that you can't detect the change, with a slightly different view for each eye so that you see in 3-D. What's more, the glasses are sensitive to head movements; the view changes as you turn your head, so that you see what you would see if you were actually looking around there. And you can be *anywhere*.

Well, almost anywhere. You can be, that is, anywhere for which the appropriate images are stored in the computer. But those scenes can be of a kind that you are never likely to see in person; at present they include for example, childbirth, seen from the inside, and the heart of a nuclear explosion.

NOT ALL OF the Sandia National Laboratory, on the hills outside of Albuquerque, New Mexico, is readily available to outside scrutiny. In fact, there is a good-sized chunk of Sandia that most of us have no reasonable hope of getting into at all, unless we are gifted with some hefty security clearances. That half of Sandia is devoted to research on nuclear weaponry and related ways of killing people en masse.

The kinder, gentler half of Sandia, however, concentrates on solar-power research, and there's nothing classified about that. Sandia isn't the only solar-power research facility in the country, of course. In fact, it isn't even the center; the Department of Energy solar-power headquarters is in the town of Golden, Colorado, in the hills outside of Denver. But Golden is primarily the place where the administrators and theoreticians live and work; there's not much to see there.

Most of the actual physical experimentation on solar power takes place at Sandia. Up in those hills the Sandia scientists have constructed horizontal- and vertical-axis windmills of many designs; there they have long strips of photovoltaic accumulators, the layered films that catch the photons from sunlight and turn them into the electrons that make electric current. There, some years ago, they built the world's first Solar Power Tower.

One way you can get electrical power from sunlight, of course, is to use the Sun's heat to boil water into steam, then run the steam into a turbine generator. That's simple in concept but not at all easy to do in practice . . . as we all know from the observed fact that we can put a bucket of water out in our backyard in full sunlight in July and leave it there as long as we like, but it will never boil into steam.

The problem is not that there isn't enough heat. There's certainly *plenty* of heat in sunlight—the amount that falls on the Earth every fifteen minutes is enough to fuel all the world's power stations for a year—but it is not readily useful heat. It comes in a highly diluted form, it is subject to interruption when clouds cover the Sun, and it is totally out of service for just about half of every twenty-four-hour day, because the rotation of the Earth cuts off the sunlight during the phenomenon we know as "night."

The first step, then, is to concentrate that solar heat, whenever it is available, so as to boil water with it. You can do that with lenses or concave mirrors, but those are expensive to build and easy to damage. A better way is to use ordinary flat mirrors, but a lot of them, all so positioned that every mirror reflects its sunlight onto a single, small common target. When you do that you can get that target plenty hot enough to boil water or for almost any other purpose you can imagine—temperatures up to several thousand degrees have been achieved, limited in practice only by the number of plane mirrors you want to assemble to direct their rays on that single focus.

That's all straightforward enough. Where it begins to get hairy lies in the fact that the Sun does not stay in one point in the sky. It rises, crosses the meridian, and sets. For that reason you have to put motors on each one of those mirrors so that it will follow the Sun and calculate just what angle they must assume with it (the angle constantly changing with the time of day and the time of year, and with allowance for the geometrical problem of having all the individual rays converge on that single target.)

But when you do all that you have something like the Solar Power Tower. When my friend Jack Williamson and I climbed it, years ago, it was taller than a ten-story building, the highest point around, with a grand view of the mountains, the labs, and the distant city of Albuquerque.

What it was not doing, however, was generating any electricity.

Well, its tenders told us, it wasn't supposed to, not at that point. It was a proof-of-process model, built only to show that the concept worked. The Sandia tower did not even have a generating system built into it. In fact, the scientists were not yet decided on whether they wanted to use that vast heat to boil steam in the conventional way or instead to convert it directly into electricity through the high-tech (but technically difficult) process called magnetohydrodynamics, or MHD. In any case, plans were already made to build a *real* Solar Power Tower in the California desert, and that one would, it was promised, actually generate real megawatts of electricity, which would go right into the transmission lines of Pacific Gas and Electric, winding up to run toasters and television sets all over the state.

It all happened just as they promised . . . for a while.

The real tower was built, out in the Mojave Desert. It worked. It generated its electricity, exactly as it was supposed to. And after Pacific Gas and Electric had operated it for a couple of years, they shut it down. It just didn't pay to keep it going, they said.

YOU MAY THINK that statement improbable. After all, by definition solar energy is *free*; the Sun keeps pouring it out endlessly. How, you ask, can it pay better to go on spending money to purchase the coal or oil to burn in a fossil-fuel plant than simply to use the Sun's freely given warmth?

There are two answers to that. The first is that the Mojave Power Tower did cost money to operate. The sunlight was free enough, but maintaining the complicated rotating mounts for its mirrors and the even more complex systems that turned the collected heat into electricity cost money. It cost more than one would expect, in fact, because the Mojave tower was only a second-generation machine, not yet fine-tuned to engineering perfection. Breakdowns were fairly frequent, as they would have been with any new technology—as, in fact, they still are with many of the world's nuclear-power plants, for instance—and keeping all the parts in repair was expensive. And all those mirrors accumulated dust and grime, so they had to be cleaned frequently.

The other reason is simpler. It is just that the bookkeeping was false. Like most such comparisons involving renewable-energy sources, the figures that measured solar vs. fossil-fuel expenses did not take into account the true costs of fossil fuels, everything from pollution and their contribution to global warming to the cost of maintaining a fleet in the Persian Gulf to protect our oil supplies.

The Mojave Power Tower isn't going completely to waste, though. In 1991 it was put back in service—not to generate electricity but to supply heat for a fish farm.

Meanwhile, Sandia is working on new and better solar-power technologies. And, oh, yes, if you do get a chance to go to Sandia, there's one thing you should not miss. Although almost everything connected with nuclear weapons is still kept under high-security wraps, there is a small and unclassified museum of nuclear weapons near the entrance, where you can see the actual bombs and missiles (the nuclear charge removed) of every variety that the U.S. arsenal has ever possessed.

THE NATIONAL LABORATORIES aren't the only places where exciting science is being done in the United States. There's plenty of it going on in universities and private industry. The difference is that private labs don't count on tax money to survive, so they do not have the same incentive to open their doors to the casual sightseer.

Still, many of them do it from time to time anyway as good public relations. It was in the RCA proprietary laboratory in New Jersey that I saw my first hologram long ago. It was almost as long ago, at General Motors in Michigan, that I saw the test tracks where they were experimenting with computer-driven cars. (That one they gave up on.) I've also had interesting visits to the Toyota main plant in Osaka, Japan; to Motorola just a mile from where I live, in Schaumburg, Illinois; to Hewlett-Packard, in California, where I was fascinated to see how they floated calculator components on streams of liquid mercury from point to point; to Bell Labs, in Holmdel, New Jersey, when it was still Bell Labs and one of the brightest jewels in America's research community; to the veterinary research laboratory at Texas A&M University, with its displays of animals with two heads, extra limbs, or no limbs at all (that lab actually did smell a bit odd); and to many more.

Unfortunately, there's no good general rule for obtaining admission to such places. One way is to watch for announcements of an open house, usually in connection with some sort of anniversary or special occasion; another is to know someone who works there. So that sort of thing is not quite as simple as visiting a national laboratory . . . but it's worth the effort!

I SAID THERE wasn't much fun in watching scientists think, but there are a few places in the world where some very great scientists did do some very productive thinking. For instance, Princeton University has kept Einstein's study just as he left it, complete with blackboard and pipe. In England, Charles Darwin's home has just now been turned into a museum, run by London's Natural History Museum (to which you can apply for details). I haven't actually seen Darwin's house for myself—yet—but it's high on my list of things to do next time I'm in England, if only to stroll the woodsy path he walked while pondering questions of evolution and speciation as he was writing his great books.

Then there are a couple of spots that I find interesting, though there's nothing much to see. For example, we know the exact spot where Leo Szilard got the idea that led to the atomic bomb. There isn't even a plaque to mark it, but it happened in 1938, while he was waiting for a traffic light to change on London's Southampton Row. Szilard had been remembering H. G. Wells's old science-fiction novel about atomic power, *The World Set Free* and had been reading about the nuclear-fission experiment of Otto Hahn and Lise Meitner, and the lightbulb went on over his head. If you could find an element that fissioned spontaneously, and in the process released enough neutrons to make other nearby atoms fission, he reasoned, why, then you might have a chain reaction that would release a lot of energy. And so, in Stagg Field a couple of years later, it happened.

(There's another Szilard site that I find personally interesting. If you walk past the old Webster Hotel at 40 West 45th Street in New York City, you might be interested to know that Szilard lived there for a while, on his way to Chicago's Stagg Field and that first working nuclear pile. All right, that's not very exciting to the general public. What is of most personal interest to me about it is that I once lived in the same place.)

* * *

OF COURSE, SCIENCE is not merely what happens in laboratories. There's a great deal of it, of many varieties, all around us, all the time. For some of the most interesting kinds we don't have to travel to some laboratory. We don't even have to leave our own back-yards . . . as we'll see in the next chapter.

Chapter 2
STARING INTO SPACE

Astronomy in Your Backyard

I suppose it is a child's natural state to be both curious and ignorant. At ten, I think I was more of both than normal, particularly when it came to figuring out just what it was that I was seeing when I looked at the nighttime sky. It wasn't that I didn't know the names. I knew all the names full well—planets Mars, Venus, Jupiter, even the then-newfound Pluto; as well as stars: Arcturus, Sirius, Aldebaran, Polaris, and others. How could I not, when I was reading every scrap of science fiction I could get my hands on, and every issue of every magazine was full of stories of Earthmen jumping into spaceships and visiting all those places? What I didn't know, what I hadn't a clue about, was which names belonged to which objects.

I don't imagine there's a ten-year-old alive in America today who is quite as ignorant of astronomy as I was in 1930. (Thirty-year-olds, yes. The country is full of college graduates who received their sheepskins with a sigh of glad relief, since that meant they would never have to know anything about school subjects again as long as they lived.) But for kids it's different now. Now it is not possible to own a television set without being exposed to *some* lore about the planets of our solar system; they're on the news whenever a spacecraft flies by to take their pictures, and over and over again on documentaries. Even the brain-rotting TV cartoon fare we offer children on Saturday mornings occasionally will show some more or less realistic picture of the planets; and, of course, hardly any American child is deprived of the class trip to the nearest planetarium.

It wasn't like that when I was ten. There was no television. Carl Zeiss's great invention of the planetarium projector had not yet come to America. There was no reliable science reporting in most newspapers, either; and although there were already a number of established popular-scientific magazines that contained all the information I could have wished for, I had never heard of any of them.

Then, when I was twelve, capricious Fortune granted me a wish I hadn't known enough to ask for. An adult friend of the family who owned a season subscription to the Brooklyn Academy of Music came to the house one day and told me that the academy offered a lecture series on astronomy, with telescopes brought onto the roof of the academy building for subscribers to look through when the skies were clear. He invited me to join him there; my parents allowed it; I was thrilled. When we got there I put my eye to the eyepiece of the old twelve-inch refractor the lecturer had brought with him and there before me lay . . . Mars. The real thing.

It wasn't red, exactly, but more a sort of yellow-orange. But all the same it was conspicuously and inarguably a real *planet*. Not just a point in the sky, but an actual disk that looked as large in the telescope as ever the Moon did to my naked, squinting eye.

It was a great experience. Like most of the great experiences of my life, the principal effect it had on me was to make me want more.

BY AND BY, as I began to learn a little more about those bright things up in the sky, I was able to pick out two more planets. They weren't that hard to find, once I found out what to look for. The easiest was Venus, brighter than anything else in the sky when it is visible at all, and usually fairly near the horizon in evening or early morning. The other was Jupiter, almost as bright, and identifiable by the fact that if the thing that shone so intensely was too high in the sky to be Venus, then it had to be Jupiter.

But if I had made a start on the planets, I had no clue which star was which. I couldn't even recognize the constellations. The only one I was sure of was the Big Dipper; that one, at least, did look a little bit like the tin dipper we drank from at the pump on my uncle's Pennsylvania farm. However, it turned out that that wasn't the thing's proper name, anyway. It was officially called Ursa Major—meaning "the great bear"—although it didn't resemble any bear I had ever seen. (Later I found out I wasn't the only person who didn't see a resemblance. The English didn't think it looked like either dipper or bear; they thought it was more like a wagon, and so they called it King Charles's wain; it didn't look anything like a wagon to me, either.)

What I learned from this is that to any rational observer (by which

I mean you and me), none of the constellations look a bit like the things they are supposed to represent. No matter how inventively you connect the dots, Sagittarius does not look like an archer, there is no visible fish in Pisces, and Libra does not resemble a set of scales.

These fanciful constellation names were given to them by ancient Greek astrologers—not astronomers; the ancients gazed at the stars for mystical rather than scientific reasons—and they had vivid imaginations. They supposed that the constellation of Andromeda looked like a woman lying down, heaven knows why. So they gave it the name of the legendary Princess Andromeda, who was supposed to be the daughter of King Cepheus and Queen Cassiopeia. The princess was beautiful, and knew it. She claimed to be even more beautiful than the Nereids, daughters of the sea god; who took offense at her presumption and sent a monster named Cetus to destroy her. Andromeda's parents were afraid of the monster, so they chained their daughter to the shore so the monster could get at her without messing up their kingdom; and she was rescued in the nick of time by the winged horse, Pegasus.

Why am I telling you all this? Only because all these mythical figures—Andromeda, Cassiopeia, Cepheus, Cetus, and Pegasus—have constellations named after them, and they're all clustered around the constellation of Andromeda. The Greeks didn't just like to look at the stars. They wanted them to tell a good story.

I knew none of this, and I sought in vain for patterns of stars that looked anything like the constellation names I had learned. They weren't there. There was only one group of stars that I came to recognize pretty quickly, because it was so easy. It wasn't anything as grand as a constellation; it was what I now know enough to call an asterism (a group of stars not important enough to be called a constellation), and it certainly was conspicuous. It consisted of three pretty bright stars arranged in a line, and it was there in the sky a lot of the time. (In fact, for most of the year it is visible almost everywhere in the world at some time of the night.) I got familiar enough with it to give it a name. I called it "Three in a Line."

My childhood "Three in a Line" is still there, and still conspicuous. If you go out and peer up at the heavens (winter is the best time to look for it, because then it's high in the sky), you can spot it as easily as I did. And then you can know, as I did not, that it is a part of the spectacular constellation which is called Orion, "the giant hunter" (and pronounced like "O'Ryan"). Those three lined-up stars are Orion's belt.

Orion is a constellation worth knowing, because you can use it to find a good many interesting objects. If you buy in to the notion that my three-stars-in-a-line are the belt Orion wears, you can then go looking for the rest of the giant hunter the Greeks thought they saw there. (Here, too, other people thought they saw quite different pictures in the same stars. The Egyptians thought Orion looked like their god Osiris, while some South American tribes were pretty sure it resembled a criminal who had been thrown down to be eaten by buzzards.)

The first things to look for in Orion are some of his other body parts. You will notice four bright stars arranged around the belt, at some distance. The brightest of them is said to be orangey-red in color (though to tell the truth I find star colors really hard to recognize), and it is called Betelgeuse (pronouncing it as "Beetlejuice" is close enough—and now you know where the comic demon in the movie got his name). Betelgeuse is supposed to be the giant's right shoulder. His left shoulder is another bright star, Bellatrix; his left knee is Rigel (RYE-jell) and his right knee Saiph (SIFE). The stars in his belt have names, too—Mintaka, Alnilam, and Alnitak—but there's no reason to know this except to show off for your friends.

You may wonder why all these star names sound sort of Arabic. That's because they are. The names were given to the stars by ancient Arab herders, who had little to do at night except to stare at the heavens.

I'm not going to try to identify all the stars in the sky for you. (If you want to pursue the matter, any bookstore can supply you with a handy, pocket-size volume of star charts.) But Orion is a good place to start learning the skies. Orion's belt is a pointer toward Sirius, the brightest star (not planet!) in the sky. Just follow the line of the belt to the left until you come to a really bright one, and that's Sirius. Orion can also show you the way to Procyon (pro-SIGH-un), Aldebaran, and the pretty little naked-eye asterism called the Pleiades (PLEE-uh-deez).

And there are two features of Orion that you may find more personally interesting, because when you look at them you are looking at the life history of our own Sun. One of them is an image of our Sun's billions-of-years-away far future, the other of its equally remote past.

The thing about our Sun is that it's a middle-aged star. It still has plenty of hydrogen left to fuse into helium, thus releasing the nuclear energy we see as sunlight. But sooner or later all that hydrogen will be used up. Then the Sun will undergo some serious changes, starting with becoming a red giant.

That's billions of years in the future. We won't be alive to see it, and actually if we were the red-giant phase would kill us all anyway. The Sun will expand so much that Earth will actually be inside the Sun's outer shell, which means there won't be much left of the Earth.

The point is that that's what Betelgeuse is now, a red giant, some five hundred times the diameter of our Sun—just as our Sun will be then.

The other thing Orion can show us that illustrates part of our Sun's life history isn't a star. It's what is called the Orion Nebula.

To find it, you look at Orion's belt. You will see descending from it, at an angle toward your left, a slightly less regular line of slightly less bright stars. This is called Orion's sword, and the third object down the length of the sword, which may look reddish to you and if your eyes are good may even look a bit fuzzy, isn't a star at all. It's the nebula.

Though you can't make this out with the naked eye, the Orion Nebula is a huge cloud of interstellar gas and dust. The universe is full of such clouds, and it is from them that stars are formed. Probably it takes some sort of shock—a collision, or a nearby supernova explosion—to make the gases start collapsing. But as they do, their own increasing gravitational attraction takes over. The cloud becomes lumpy, like the raisins in a rice pudding, and as the lumps grow larger their gravity pulls more and more of the surrounding matter into their own growing masses. When enough matter has been accumulated, the density of the protostar increases to the point where the atoms of hydrogen (which is what the gases are mostly made of) are forced into each other. Atomic fusion begins, just as in an H-bomb, except that here the explosion keeps going on. Energy is liberated, and it is that energy that makes stars shine.

That's what is happening in the Orion Nebula. Stars are being born there, just as our own Sun was more than 4 billion years ago. You can't see those individual baby stars with the naked eye. It takes a telescope even to see the nebula as a patch of glowing white, and a *big* telescope to pick out the individual stars within it. And if you use a *really* good telescope—say, the orbiting Hubble Space Telescope—you can see something even more interesting.

The wonderful thing about these stellar babies, which nobody knew until the Hubble turned its eye that way in the early 1990s, is that many of them are surrounded by a disk of gas. That faint and fuzzy disk is almost certainly the same sort of material that surrounded our own Sun when it was new—and that ultimately condensed into all the planets and satellites and comets and asteroids that surround our own Sun today. So what the Hubble sees is not just the infancy of stars like our own but the birthing of whole planetary systems.

DOES THIS MEAN that every star in the sky has its own family of planets in orbit around it?

Not at all. For one reason or another, many stars are extremely unlikely to have planets. But we know now that at least some stars do, because astronomers have now found at least fifty of them.

That's not easy to do. As of this writing, no telescope has ever shown an image of an extrasolar (not orbiting our Sun) planet; by definition they can be found only close to a star, which is so much brighter that the faint image of the planet is rendered invisible to us.

But there is more than one way to skin a cat. Astronomers realized that gravitational attraction is a two-way street. Not only are planets attracted to their star—that's what keeps us from drifting away into interstellar space—but stars are attracted to their planets, as well. That is, the star is tugged a little way this way and that as planets move around it. Unfortunately, since the star is vastly more massive than its planets, the motion is very slight and hard to detect.

But not quite impossible.

As telescopes got better and sky charts more accurate, astronomers began to hope that they might find evidence of planets by carefully monitoring the position of suspect stars in space. If a star had planets—big ones—it would wobble a bit, just as Jupiter pulls our own Sun ever so slightly in its direction wherever it is in orbit. It wasn't long until some astronomers (out of charity I won't name them) excitedly reported that, yes, they had detected just such motions and thus had established that there were indeed planets orbiting around certain other stars.

It was a thrilling report. Unfortunately, it was wrong. The photographic plate holders in the observatory responsible for the reports had been moved while the observations were going on; it was the movement of plate holders, not the stars, that had been detected.

Scratch that idea, start over with a better one. There is a better way of identifying small motions in a distant star: the spectroscope. The color of light changes with the motion of the light-emitting body.

If it is moving away from us, the light is shifted toward the red end of the spectrum; toward us, toward the blue end. These, too, are tiny, tiny shifts, but it's easier to measure color shifts in a spectrum than the gross motion of the stars themselves. And in that way, since 1965, more than a dozen stars have been shown to have planets.

You aren't going to see those planets for yourself. You can do the next best thing, though. You can see at least three of those planetiferous stars, and you can do it with the naked eye.

The easiest to find is the star called 47 Ursae Majoris. The two stars on the handle side of the Big Dipper's bowl point downward to it. Not one of the really bright stars, 47 Ursae Majoris is fifth magnitude. (The faintest stars you can hope to see with the naked eye are designated sixth magnitude. Fifth magnitude isn't that much brighter.) That means that you need fairly good seeing to pick it out, but there it is. If you imagine the Dipper's handle bent back along the bowl, its tip would be quite close to 47 Ursae Majoris. Other fifth-magnitude stars now known to possess planets are 51 Pegasi (just above the Great Square in Pegasus) and 70 Virginis, in the constellation of Virgo but not easy to locate without a good star map.

SO FAR WE'VE been looking at relatively nearby objects in the vast scale of the universe. Even the Orion Nebula is only about 500 light-years away, and most of the stars I've mentioned are a good deal closer. (A light-year, of course, is the distance light travels, at 186,000 miles per second, in the course of a year—186,000 × 60 × 60 × 24 × 365. You can do that arithmetic for yourself if you like, but the resultant number is too big to mean much. Settle for realizing that that's a lot of miles.)

Perhaps by now you'd like to get a look at something a good deal farther away. For that, you want to look at an external galaxy.

In case you don't know the difference between a galaxy and a solar system, a solar system is only a single star and the objects, such as planets and comets, that revolve around it; a galaxy is a large number of stars, usually upward of 100 billion of them, held together by their common gravity. (If you know that much you already know more than a lot of newscasters and almost all writers of sci-fi television programs.)

Those external galaxies—galaxies outside our own Milky Way Galaxy—are very far away from us, some so distant that the light we see from them in our biggest telescopes began its journey not long after the universe itself was formed, more than 8 billion years ago.

However, there are plenty that are closer than that. There are even a few that you can see with the naked eye.

You won't, however, be able to see the most impressive of them if you are very far north of the equator, because they are Southern Hemisphere objects. These are the two Magellanic Clouds, so called because the first Europeans who saw—and named—them were sailors on Magellan's historic first round-the-world voyage a few centuries ago. The names are appropriate, because they do look like faint, high clouds in the sky. As galaxies go, the Magellanic Clouds are neither particularly well formed nor large. They are, however, two small galaxies that are captive to the gravity of our own, much larger galaxy; they revolve around our own galaxy in the same way that the Earth revolves around the Sun (though, of course, much more slowly), and probably at some very remote future date they may be drawn right into our own galaxy and disappear as independent objects.

The other fairly easily observed external galaxy goes by the astronomical title of M–31, more familiarly known as the Great Nebula in Andromeda.

The best time to see the Andromeda Galaxy is October, but it's reasonably high in the sky any time from Labor Day to Christmas. Look toward the north on any clear night. You'll see a sort of wobbly letter *W* in the sky; that's the constellation Cassiopeia. Below Cassiopeia and to the right is another reasonably easily spotted asterism. (Remember that an asterism is a group of stars not dignified by the name "constellation.") This one is composed of four pretty bright stars forming the corners of a large square. The constellation these four stars are located in is a little larger than Cassiopeia and it is called Pegasus, so the four stars we're talking about are called the Great Square in Pegasus.

Having spotted these objects, draw an imaginary line from the left-hand side of Cassiopeia to the upper left-hand star in the Great Square, which is called Alpheratz. Then, about two-thirds of the way down from Cassiopeia to Alpheratz, and a little to the left, you should be able to make out a small, fuzzy, white blob.

That is M–31, the Great Nebula in Andromeda, and what you are seeing is the collective light shed by its 800 billion or so stars—several times as many as we have in our own galaxy. That's a lot of stars. They give off a lot of light, which is fortunate for us amateur stargazers because it takes that much light to make M–31 visible to the naked eye—it is an object of 4.8 magnitude. And it is a little more than 2 million light-years away.

That sounds like a great distance. In human terms, it is. But the distances astronomers deal in are so much greater than that, that M–31, along with the Magellanic Clouds, our own galaxy, and about thirty others, great and small, is considered to be a member of what is called the Local Group. What unites this Local Group is not merely that they are (relatively) close together. They are also held together by their common gravity. Each one of the thirty-odd galaxies in our group is in some sort of complicated, slow orbit mediated by the attraction of all the others. Such groups of galaxies are called clusters. The sky is full of such clusters, some of them with many more members than our Local Group.

Gaze at M–31 with respect; it is the most distant object any human being can see with the naked eye. (And if anyone ever asks you how far you can see, you can truthfully say "Oh, about 2 million light-years.")

So MUCH FOR the external galaxies. Now let's talk about the one we live in, the Milky Way.

Most books on astronomy will tell you that, under the best of conditions, the most stars you can see with the naked eye is about four thousand—and only half of that number at any one time, because the other two thousand are on the other side of Earth.

That's not exactly true. If you look in the right place you can see many, many millions of stars. The trouble is that there are so many of them, and so apparently close together, that they all blend into a faint, pale, silvery cloud that looks quite a lot as though someone had spilled a cup of milk across the sky . . . which, of course, is exactly why our galaxy is called the Milky Way.

Although only a fraction the size of M–31 in Andromeda, our Milky Way is still a big one, as galaxies go. It contains somewhere around 300 billion stars—give or take 100 billion or so; different astronomers give different estimates. Some of those hundreds of billions of stars are far bigger and brighter than our own Sun, though most are smaller and dimmer, and, like M–31 (but unlike the amorphous-shaped Magellanic Clouds), the Milky Way is a spiral galaxy.

This means that if you could look down on our galaxy from above one of its poles, it would look like a giant pinwheel. In a spiral galaxy, swirls of stars and masses of gas and dust spin out from its center. The particular swirl our Sun is in is called the Orion arm. Nearly all the stars you can see overhead are also in the Orion arm, though a few dimmer objects are in the adjacent Perseus and Sagittarius arms—Sagittarius closer to the center of the galaxy than our own arm, Perseus farther out.

When you look at that glowing Milky Way in the sky, you are looking toward the center of our galaxy. You can't see very far into that milky spread, though; all that brightness defeats your eyes. What lies beyond that spread of blended starlight is the core of our galaxy.

Even the biggest optical telescopes—the ones that see by the same sort of light we do—can't see to the core, but fortunately astronomers have instruments that can see in other, nonoptical frequencies. What these instruments reveal is that the core is a bizarre place, where many, many stars and gas clouds are squeezed close together and seem to share the neighborhood with a giant black hole. In the core, all these sources produce a vast wash of radiation, not just visible light but deadlier kinds. If you were standing in the core to look outwards, you would see a spectacular sky, dense with stars, and many of them far brighter than anything visible from Earth. But you wouldn't see it for long. The radiation would kill you.

And beyond the core, hidden even better from us, is the other half of our own galaxy.

HAVE YOU HAD enough of objects that are hard to see? Let's talk about some that are exceedingly visible, our own Sun and Moon.

What we all know about the Sun is that it's easily the brightest thing around. It looks as though it moves across the sky from daybreak to dusk. People used to think it actually did, but now we know it just looks that way. The Sun stays (relatively) still, while our planet rotates under it. And we know that the Sun always looks pretty much the same to us.

The reason for that, though, is that the Sun is so painfully bright that we can't really get a good look at it with the naked eye. What we need to look at the Sun is something that will cut out most of that blinding—literally blinding—light. Ordinary sunglasses won't do. You need to wear something a lot stronger; welders' glasses with a protection rating of no less than 15 are recommended. And if you do that, you'll probably see that the Sun is not a featureless disk after all. It is mottled with dark spots.

These sunspots aren't really dark; they're just not as bright as the surface around them, and they seem to have an effect on Earth's weather. There was a period a few centuries ago when the sunspots were so scarce that many naked-eye astronomers never saw one at all and didn't really believe the reports of them in the old literature. That period is called the Maunder Minimum, and it was associated with long stretches of very cold winters on Earth.

To look at the Moon you need no such protection. All you need is the right time of month (the word "month" derives from the word "Moon") and a sky free enough of clouds for the Moon to shine through.

What we observe first about the Moon is that sometimes it's full and round, sometimes only a crescent, and some nights it isn't there at all. You probably know why that is. But in case you don't, or if you want to explain it to someone else, the reasons are easy to demonstrate. Put a few coins on a tabletop. Let a quarter represent the Sun, a dime the Moon, a nickel Earth. (The proportions are all wrong; the Sun is relatively much bigger than that, and much farther away than your tabletop will allow, but for this purpose close enough. Move the dime around the nickel, as the Moon moves around the Earth. Remember that the Moon is always "full" if viewed from the Sun—that is, the hemisphere facing the Sun is always fully illuminated by sunlight. You can easily see that when the Earth is between the Moon and the Sun, we are looking at that fully lighted hemisphere; when the Moon is between the Sun and the Earth, we're looking at the dark hemisphere, and consequently can't see it at all; and at other times we are seeing only that portion of the Moon's surface that is receiving sunlight.

That's all very simple, but it's astonishing how few people actually know that. Even some bright ones didn't, as for instance Samuel Taylor Coleridge, the very intelligent man who wrote the poems "Xanadu" and "The Rime of the Ancient Mariner." In the "Rime" he speaks of

> the horned Moon, with one bright star
> Within the nether tip.

You'd think Coleridge would have known that the dark part of the Moon doesn't disappear so that you can see stars through where it used to be. But he didn't.

The best times for observing the Moon are somewhere between new and full. Full moon is pretty, but the sunlight on its surface comes straight down; that means you can't see the shadows that reveal most of the detail.

* * *

IF YOU LIKE to ponder strange and unexplained coincidences, here's one for you:

As seen from Earth, the Sun and the Moon appear to be almost exactly the same size.

There isn't any astronomical reason why we should have such a satellite. It isn't because of some vast physical law; no other planet in our own solar system is similarly gifted. Mercury and Venus don't have any moons at all, and, while the rest of our Sun's planets do—sometimes more than a dozen of them—not one of them shares that coincidence with Earth. The coincidence of apparent size is just a curiosity, but it's a very lovely one for us Earthlings because it produces the best eclipses in the known universe.

Eclipses of the Moon occur only when it is full—that is, when Earth is between the Sun and the Moon—and the reason the Moon is eclipsed is that it moves into the shadow cast by Earth. You may wonder why that doesn't happen every time there's a full Moon. The answer is that the plane in which the Moon goes around Earth isn't quite the same as the plane of Earth's revolution around the Sun. Most of the time when the Moon is full, it is either a little above or a little below Earth's shadow. Sometimes it is pretty close, and you get a partial eclipse as the shadow blacks out some fraction of the Moon's surface. But it's best when the eclipse is total and then you can see that the eclipsed Moon isn't totally dark; it's faintly, reddishly visible, because it is picking up a little bit of light refracted from our planet's atmosphere.

The good thing about a lunar eclipse is that whenever one occurs, it is visible everywhere on the Earth where the Moon is in the sky at all.

Not so for solar eclipses. The Moon is a lot smaller than Earth. It casts a much smaller shadow. When, in its new-moon phase, the Moon gets between the Earth and the Sun, the shadow it casts is only a few miles wide. Only those fortunate enough to be in that tiny patch of the Earth's surface can see the eclipse in its totality. (Those just outside the shadow see a partial eclipse—striking, yes, but nothing like the real thing.)

What makes the situation more difficult is that the Earth is rotating under that shadow. (The Moon is moving, too, though much more slowly.) The result is that the shadow doesn't stay in the same place for very long; it makes a track somewhere across the Earth's surface for a thousand miles or more and is never visible in any one spot for more than a maximum of seven minutes—sometimes only about four.

But, oh, my friends, what a wonderful few minutes those are!

I've been lucky enough to see two total eclipses in my life. The first (I found out later) wasn't *absolutely* total; I was a few miles out of the zone, and so it was only about 99 percent total, but it was enough to awe the six-year-old Brooklynite I was at the time. The second, though, was the real thing. I saw it from the cruise ship *Independence*, off the coast of the Big Island of Hawaii.

One of the thrills of solar eclipse–watching is the suspense: Are you going to see it at all, or will clouds move in just when you don't want them and you see nothing? We had that problem on the *Independence*. Clouds were all over that part of the Pacific at just the wrong time, leaving only a few clear patches here and there. Fortunately we had a weatherman on board, Joe Rao, and he was in touch with all the satellite data; he could tell the captain where the nearest clear spots were, and the captain wheeled that great liner around the ocean like a sports car until we found the right place at the right time. Our view was perfect.

Most things you can see on television at least as well as if you are there in the flesh, sometimes much better. That's not true of a total eclipse of the Sun. What the camera can't show you is the way that bizarre object over your head, a dead-black circle surrounded by the licking flames called prominences, dominates the entire sky. Pictures don't convey what it's like. Descriptions don't, either. You have to see it for yourself.

Unfortunately, that isn't always easy. Although there is a total solar eclipse just about every year, they aren't always in the same place. You may have to travel some distance to see one, but it's worth it.

To give you an idea of what I mean by "worth it," let me tell you about a young boy whose grandfather had taken him on the *Independence*'s eclipse cruise. He was not a happy traveler. There weren't any Nintendos on the ship, and he missed his friends. His grandfather tried to persuade him that this would be something to tell his children about, but the boy was having none of that; he wasn't ever going to have children, he said, so what was the point?

Then he saw the eclipse.

As we were standing on the deck, all thousand of us, shaken and exhilarated by what we had seen, I heard him telling his grandfather that the old man was right, this was something to tell the children about. "But I thought you weren't going to have any," the grandfather teased. The boy shook his head. "Now I have something to tell them," he said.

* * *

OBSERVING AN ECLIPSE isn't like simply looking at what is always there in the sky. It's an event. And there are other events that can be seen.

For instance, there are comets. A comet is a huge chunk of dirty ice. It spends most of its life in the outer reaches of the solar system, where it can't be seen. But as a comet swings in to circle about the Sun, before climbing back into the darkness, it warms up. As its frozen gases warm, they spread out in a tail that is sometimes spectacular. The tail of the Great Comet of 1811, for example, was 100 million miles long—even longer than the distance between the Earth and the Sun—and to observers on the Earth it stretched over almost half the sky.

Whether you will ever see one that impressive is a matter of luck; the most spectacular comets have come from nowhere, and no one knew they were coming until some astronomer more or less accidentally spotted them on the way. Others are more predictable—like Halley's Comet, coming back to visit us every seventy-five years or so—but by and large those usually aren't as exciting to look at.

A rarer—*far* rarer—sky event is a naked-eye supernova.

A supernova is the death of a star. It takes a big star to make a supernova; but sooner or later every really big star will run out of enough of its hydrogen fuel to collapse. Then the collapse itself triggers an immense rebound; the dying star emits thousands of times as much light as it ever had in its normal existence, and if it is close enough—it doesn't have to be *very* close—we see it as a new star in the sky.

However, that doesn't happen very often. When I say naked-eye supernovae are rare, I mean *really* infrequent. The last good one in our own galaxy was in 1572. When there will be another no one can say, but they are so bright that sometimes we can see one that isn't in our own galaxy at all.

That happened in 1987, when an astronomer in New Zealand happened to glance up at the sky one night and see a star that didn't belong there. It wasn't in our galaxy. It was in one of those nearby Magellanic Clouds, and it was so intrinsically bright that, even at that distance, it was clearly visible to a human's naked eye.

How bright was that, exactly?

Well, to figure that out we need to do a little arithmetic. A distant object, we all know, is fainter than one with equal light nearby. The rate at which it gets dimmer depends on the square of the distances.

(This is called the law of inverse squares.) A candle that is twice as far away as some other candle is dimmer than the nearby one by the square of 2 (which is 4) divided into 1; that is, it's only one-quarter as bright. So for SN1987a (as that Magellanic Cloud supernova is called) to appear as bright as some nearby normal star, its absolute brightness has to have been the square of the difference between a few hundred and a few hundred thousand light-years—call it at least 100 million times as bright.

That's *really* bright. You might ask what would happen if a supernova like that were to occur somewhere nearby, instead of all that far away.

At least one scientist at the University of Chicago thinks he knows the answer to that. He believes exactly that did happen once, a very long time ago. What makes him believe that is that all over the Earth are layers of rock which show traces of minerals that appear to have been deposited here by just such a supernova explosion. They are all in the same rock strata, dated at about 230 million years ago.

That's an important date in the history of life on Earth. It is the time of what is called the Cambrian extinction, when, at the end of the Cambrian age, a majority of all then living species went extinct. Nobody knows for sure what caused that great die-off, but the supernova theory is as likely as any other.

If it happened once, can it happen again?

Oh, yes, at some time in the long future history of the Earth. However, nobody knows when that time may be. It could be the day after tomorrow, or it could be many billions of years from now. So if you're a worrying kind, there are more urgent things to worry about that; meanwhile, if another supernova does appear in the sky, it certainly would be something worth looking at.

A MORE BENIGN, and far more common, event in the sky worth a look is a meteor shower.

Meteors—sometimes inaccurately called shooting stars—are bits of matter that fall onto the Earth's atmosphere. Most of them aren't much bigger than a sand grain, but they are moving so fast that the friction of the air heats them to incandescence, and we see them as bright lights moving across the sky. Most meteors burn up in the air, but now and then a particularly large one lasts long enough to land on Earth's surface (at which point it isn't called a meteor any more but a meteorite).

The best time to see meteors is after midnight, because then they are colliding with our planet head-on. The morning side of the Earth, the side that you are on from midnight to noon local time, is the Earth's leading edge as it moves around the Sun. So there are more meteors then, just as when you are hurrying to get out of a sudden shower you'll get more raindrops on your face than on the back of your neck.

If you're patient, you may see a meteor almost any night of the year. But if you want to be pretty sure of a show, you want to be out there looking when a meteor shower is scheduled.

It's comets that create meteor showers. As a comet ages in its orbits around the Sun it begins to break up, shedding a trail of debris. When Earth in its own course passes through that trail, the meteors are more frequent, sometimes averaging one a minute, occasionally many more than that. And—a blessing for backyard stargazers who want to plan their viewing—Earth hits those cometary debris trails at the same time every year.

If you want to observe a meteor shower, you'll be looking up at the sky for some time. To avoid getting a crick in the neck, that is best done on a lawn chair or a hammock, anything so that you can lie back to watch. (My own favorite way of doing it for summertime showers, when I happened to own a house with a swimming pool, was to float on a rubber raft in the pool after the mosquitos had given up for the night.)

The great meteor showers all seem to radiate from one or another particular point in the sky, and so they are named after the constellations they seem to come from. Here are a few of the best of them:

January 3: The Quarantids. They're slow-moving so their trails stay in the sky a little longer than most; they usually put on a good show.

April 22: The Lyrids. These aren't numerous, but sometimes they are quite bright; they seem to come from near the bright star Vega.

August 11–12: The Perseids. These are the remains of the Swift-Tuttle Comet, averaging about fifty meteors per hour. Perseids are everybody's favorite meteor shower, because they come at a good time for lounging outside to watch the display.

November 17–18: The Leonids. These are remnants of the Tuttle-Temple Comet, and they move fast. How many you see depends on what part of the stream we are hitting on that particular night. The Leonids are on a thirty-three-year cycle; from its high point in 1999, the display diminishes until 2016, then it increases again to its next high point in 2032. The years 1833 and 1966 were vintage years, with displays that some observers compared to the flakes of snow in a storm. In a lean year, however, the Leonids may produce only half a dozen meteors in an hour.

December 13: The Geminids. These peak at about sixty an hour. They strike the atmosphere at only about seventeen miles per second, half the velocity of the Perseids, so they seem to linger longer in the sky. They might rival the Perseids for popularity, if it weren't for the fact that December in the Northern Hemisphere is not as attractive a month as August for staying out of doors to watch the sky.

The meteors we have been talking about are the bits of cosmic junk that burn up in the air. What about the ones that don't burn up?

Mostly they simply fall to the ground, doing no particular harm. True, once in a great while one hits a house, a car, or even a person—that last has been known to happen just twice, out of all the billions of people there have been, so don't lose any sleep over it. If you want to see what one looks like, your local museum probably has a selection—the new Rose Institute of Earth and Space Science, which recently opened on the site of the Hayden Planetarium of the American Museum of Natural History in New York, displays one about the size of a Buick, laboriously brought down from the Arctic. Most are much smaller.

But big ones do strike the Earth from time to time, and some of them leave their mark. In Barringer, Arizona, one fell 49,000 years ago and scooped out a hole in the ground a mile wide; that's called Meteor Crater, and it is still there to see. That's not the biggest. The lower part of Maryland's Chesapeake Bay is the remains of a metorite crater left there between 15 and 16 million years ago. Apparently the object that made that mark broke in half before it hit Earth; at least there's a crater thirty miles across in Popogai, Siberia, that looks as though it was formed by the other half.

They come a lot bigger than that sometimes, too. About 2 billion years ago an object about the size of Mount Everest crashed into what is now Sudbury, Ontario. You can't see much of a crater, though. Erosion by wind and rain has long ago scoured away all surface traces, as it has done for that famous meteorite that appears to have killed off the dinosaurs 65 million years ago. That one struck near the Yucatán peninsula. The area is mostly underwater with nothing to see. What you can see—it's not all that interesting—is rocks a thousand miles away that are made of debris splashed out by that impact; I've seen places where they were supposed to be in the little Central American country of Belize, but, to tell the truth, they looked like any other kind of rocks to me. And in central Texas there are traces of debris left by the immense tsunami that the impact threw out.

Of course, if you had happened to be there, 65 million years ago, you could have seen, and heard, plenty—the ear-breaking boom of the impact, the rain of debris, the tsunami, followed by the long years of darkness as the dust in the air cut off the light of the Sun worldwide and thus brought about what is called the K-T (from Cretaceous-Tertiary) extinction. But you probably wouldn't have lived long enough to see much of it.

WHEN I TALKED about the visible stars, I left out quite a few. That's because they are the southern constellations, and you can't see them from your own backyard if you happen to live in the Northern Hemisphere.

You needn't feel too deprived about that, though. True, the southern skies have those wonderful objects, the two Magellanic Clouds, but apart from that stargazers are better off in the north. The southern skies don't have nearly as many bright stars, which means the constellations are not as striking. Even Crucis, the famous Southern Cross, looks more like a rickety child's kite than a cross.

There is one southern star that is of particular interest. That's the brightest one in the sky there, Alpha Centauri. What makes it special is that it is the closest star to our own Sun that is eyeball visible, at a distance of only four light-years. (Alpha Centauri has a companion, appropriately named Proxima Centauri, which is slightly closer, but it's too faint to be seen without magnification.)

If you happen to go south of the equator, there's at least one constellation you'll recognize. That's our old friend Orion, which sits just about on the equator and is equally visible from either hemisphere. If you see it there, though, it will look odd to you. Viewed from the south, the giant appears to be standing on his head.

WHEREVER YOU HAPPEN to be, naked-eye astronomy can be a lot of fun . . . though, I am sorry to say, maybe not quite as much fun as it was in the days of my first attempts at it.

The difference is the seeing. Skywatching is no longer quite as easy as it was sixty years ago. Our cars, our home heating systems, and our power plants and factories have seriously dirtied up the air with their particulates and pollution in general. (In the United States the maximum horizontal visibility you could hope for in my early youth ran as high as ninety miles. These days it rarely goes above fifteen or twenty.) That's not all, though; we've also added skyglow. If you are anywhere near a city (and how many Americans are not?), all those urban streetlights and neon signs make a permanent reddish orange aurora in the sky that obscures just about every object dimmer than third or fourth magnitude—indeed, there are times when it's hard to find even that brightest of fixed stars, first-magnitude Sirius. The seeing was a lot better way back when. I distinctly remember seeing the Milky Way—faint but clearly glowing—as a young teenager in the heart of Brooklyn; in recent years I don't remember seeing it at all anywhere near any large city.

Still, naked-eye astronomy has a long and honorable history. For thousands of years it was all the human race had, and yet the ancient Greeks were able to carry out some astonishing measurements: Hipparchus calculated that the Moon was 270,000 miles from Earth (the actual average value is about 230,000 miles), while in the third century B.C. Eratosthenes calculated the circumference of Earth to be 24,000 miles—just about right—and the distance to the Sun to be 92 million miles, only a bit more than 1 percent too near. These were great discoveries, and all were made with the naked eye. It's too bad that most of the human race forgot them for the next two thousand years.

All the same, the astronomer has not been born who didn't wish for a better look at what he was observing, and that means a telescope.

There are three ways of getting a telescope to use. You can make one, you can buy one, or you can borrow one from somebody else.

Making one isn't easy; it will take a lot of long, hard work, but thousands of amateurs have done it. You can order blank Pyrex mirror disks for only a few dollars each; then, at the expenditure of only a few hundred hours of hard manual labor devoted to grinding the blanks together with jeweler's rouge—and provided you were lucky enough to make your interference-fringe tests fairly accurately so you came out with a properly curved surface, and also provided that you weren't stupid enough to drop something that scarred the mirror before you finished it—why, then you would wind up with a reasonably well-configured mirror. After that it is only a matter of putting a few sections of tube together and adding an objective lens and, voilà, there's your very own telescope.

That's how you make your own. I hasten to admit that, though I've seen it done, I've never done it myself. (I almost did, once, while stationed at an Oklahoma air base in World War II, but my orders to ship out for Italy came before the blanks arrived.)

It's a lot easier to go out and buy one. The only question there is deciding how much you want to spend. As with most things, the better it is the more it will cost; there are instruments available that cost many thousands of dollars—and many thousands more for the clock drives that will keep them pointed where you want them as Earth turns, the extra eyepieces, cameras, and other accessories that give them more flexibility; and, yes, the little private dome you may wish to build to keep them out of the weather.

But you can start a lot smaller than that. My own first telescope was a child's spyglass—one of those brass, sectioned things that telescoped out to a length of a foot or so, with magnification of no more than a couple of diameters. That was some help. It was interesting to turn it on, say, the Moon, because you could pick out at least the largest of the craters, and at quarter phase you could make the fascinating discovery that the dark part of the Moon's disk wasn't entirely dark after all but faintly, reddishly illuminated by Earthshine.

In all truth, that little telescope was better than it seemed to me. It was at least as powerful as the instrument that Galileo had made for himself centuries ago—you would be hard put to find one now on sale that isn't—and he used his to stand the world of astronomy on its head by discovering the moons of Jupiter and the phases of Venus.

IF YOU CAN'T afford the telescope you'd like to have, be not discouraged. Probably you can use other people's from time to time. The world is full of astronomy clubs, organizations of people who own telescopes and get together frequently to show them off. (I'll tell you how to find them later in this book.) Many schools, and almost all universities, have small observatories and open them to the public now and then.

But if you want to see some really big telescopes, you have to go where the astronomers are. In the next chapter I'll tell you how.

Chapter 3
THE BIG EYES

Visiting Observatories

Next time you're in Boston, give yourself a little treat. Take the Red Line subway under the Charles River to Harvard Square, then a taxi or a bus to Garden Street. On a little hill over the street there you'll notice a large greenish dome set back from the street. You may think to yourself that it looks remarkably like an astronomical observatory, incongruously located right in the middle of a city.

Well, you'd be right. That's what it is. It is the dome of the giant (then) state-of-the-art telescope the Harvard College Observatory built a century and a half ago, now a part of the Harvard-Smithsonian Center for Astrophysics. Under that thirty-foot bubble of copper (already verdigrised, though it was replaced with virgin metal as recently as 1993) is the authentic fifteen-inch refractor that once was the biggest and best telescope in the United States, maybe in the world. In fact, it was generally known as the Great Refractor, just as the two-hundred-incher on Mount Palomar was affectionately called the Big Eye during its own decades of supremacy.

In its day the Cambridge Great Refractor was an astronomical powerhouse. From its "first light" in 1847 and for more than twenty years thereafter, the Great Refractor was astronomy's cutting edge. It is the instrument that discovered the eighth moon of Saturn and made the first observations of that planet's inner ring. The first astronomical photograph was taken through the Great Refractor, in 1850. Around the telescope dome, in the building that supported and served it, some of the world's greatest astronomers did some of their finest work. It is where Henrietta Leavitt pored over photographs of Cepheid variable stars in the Magellanic Clouds and from them produced the first good measurements of the distances between galaxies . . . and thus made it possible for Edwin Hubble, half a century ago, to make the astonishing discovery that the entire universe, with all its far-flung stars and galaxies, is expanding like a blown-up balloon.

That's the way it was in the middle of the nineteenth century. No more. Those days of pioneering astronomical observations are past for the Great Refractor itself. The work of studying the distant reaches of the universe has been taken over by bigger, newer, and better-located telescopes. But the instrument on Garden Street hasn't disappeared. It is still there for people like you and me to visit and admire.

The outmoding of the Garden Street telescope doesn't mean that scientific research came to a total halt at the Harvard-Smithsonian. That institution still does have a working astronomical observatory. In fact, it has several of them: There is one observatory farther out in the Massachusetts countryside, another on a mountaintop in Arizona, and a third as far away as Peru, where not only are they blessed with good seeing but the whole Southern Hemisphere of the sky, invisible to north-situated Boston, is open to their investigations.

What the Cambridge campus does now is furnish a home base for a large core of working scientists, busy with an astonishing variety of research programs. Theoretical studies, computer modeling, data analysis, and many other kinds of investigations are going all the time. It was a Harvard-Smithsonian astronomer, Gerald Hawkins, who established that some puzzling ancient megaliths like Britain's Stonehenge were actually primitive astronomical observatories. (I'll have more to say about them later.) At one time or another Harvard-Smithsonian has been the world headquarters for such diverse international enterprises as tracking space junk in orbit, coordinating the international reporting system called the Center for Short-Lived Phenomena (keeping tabs on everything from comets and volcanic eruptions to rains of frogs from the sky), and supervising the American space program's Deep Space Network of Schmidt telescopes all around the globe. All those researches and activities represent first-rate science, but they are not really a lot of fun to watch.

However, the old telescope is still there on Garden Street, and it is a beauty. Its twenty-foot shaft is veneered in mahogany. In the days when astronomical observers still peered into an eyepiece, the Great Refractor provided them with an eyepiece chair, upholstered in crimson velvet. The telescope moves on ball bearings to follow its target around the sky. Because it's so big those ball bearings had to be big, too. (Legend is that when it was first built, leftover cannonballs from the War of 1812 were pressed into service for it to revolve on, but I don't guarantee that this story is true.) And it is a refractor, which means that, unlike almost all modern instruments, it has a lens instead of a mirror.

What's the difference? Well, what any telescope does is to take the large, but diffuse, quantity of light that falls on the primary optics—which is to say, the surface of the lens or mirror involved—and concentrate it into a smaller but far brighter image at the eyepiece. The larger the lens or mirror (called the aperture), the more gathered light it has to concentrate; thus the better the image.

The way the first telescopes did that was by passing the light through a glass lens; those were the refractors. As far back as the sixteenth century, Isaac Newton showed that magnification could be done equally well with a properly configured mirror, but the refracting lens remained the optical system of choice for more than three centuries.

As a practical matter, for big instruments the mirror is much better. Optically, lenses are subject to a color distortion called chromatic aberration, because the glass bends light of different wavelengths to different degrees. Mirrors don't have that problem. Besides, in a mirror only one face instead of two has to be ground to perfect curvature, which makes it easier to manufacture. Finally, for a large instrument a mirror is easier to maneuver. Gravity is the enemy of large optical systems. It is simpler to support a large, heavy mirror against the drag of its own weight as it moves to search the sky, so it stays in perfect configuration more easily; a lens can be supported only at its edges, while a mirror's whole back is its support. The bigger the mirror, the more important that is. In fact, for some really big instruments, it has become necessary to add an array of little pistons on the back of the mirror, computer-controlled to counteract gravitational drag by pressing the distortions back into shape. You couldn't do that with a lens. A lens doesn't have a "back."

So all modern astronomical telescopes are reflectors; but a grand old instrument like the Great Refractor is still a wondrous apparatus, and, happily for us civilians, the Harvard-Smithsonian has kept it as a souvenir. The third Thursday of every month is Observatory Night in Cambridge, when the observatory is open to the public. At the moment of this writing, the Great Refractor itself is off limits because it has had a lot of wear over the years and currently is being restored to its nineteenth-century splendor. Funding has slowed the project down, but it should be completed soon. Meanwhile, there are other telescopes to look through and astronomers to listen to, and a whole lot of scientific history in the air.

WHEN THE CITY of Cambridge grew up around the Great Refractor, it doomed that splendid old telescope as far as serious research was concerned. It isn't the only instrument that has happened to; cities, with their polluted air and their myriad streetlights and neon advertising signs, are not a fit environment for astronomy. The big telescopes have fled to other climes.

If you want to see what a modern high-tech observatory looks like, you have to follow after them in their flight from human habitation. Fortunately, many of the places they have fled to are worth visiting on their own, and one such is the Harvard-Smithsonian's own Fred L. Whipple Observatory, located on the peak of beautiful Mount Hopkins, thirty-odd miles south of Tucson, Arizona.

There are several telescopes on the mountain, but the star of the show is what used to be the pioneering Multiple Mirror Telescope. The MMT wasn't just different; it *looked* different, inside and out. As you drove up the mountain, it appeared not as the half of a beachball shape of almost every other telescope in the world but as a huge brick; the whole brick revolved as the telescope scanned the heavens. Inside, as its name suggests, this marvelous instrument had not one but six—count 'em, six—primary mirrors.

What you want in a telescope is light-gathering power—a big surface aperture to collect the faint glow from a galaxy millions or even billions of light-years away—and there is no ordinance writ in heaven that says all that collecting surface has to be on a single disk. The Multiple Mirror Telescope had no fewer than six mirrors working in synchrony. Each of them was a hefty 72 inches across, and they were computer-controlled so that each one of them focused in precise alignment on the "beam combiner" that made the image. The system worked splendidly. The end result was the same as if a single 180-inch mirror were the primary optic. MMT was the first large instrument to try this complex system, but it functioned so perfectly that a number of others like it have been built since.

Unfortunately for us spectators, the Multiple Mirror Telescope is multiple no longer. It did exactly what it was designed to do, in showing that the light from multiple mirrors can be combined into a single image. But its six mirrors, though good, were not quite perfect for astronomical uses—they had been originally acquired as war surplus from the air force. So, now that it has demonstrated that the multiple-mirror system works, its six mirrors have been replaced by a single larger and lighter one.

One other great good thing about the Whipple Observatory, for the likes of you and me, is that it can be visited. In fact, the administration runs bus tours to the peak for any interested person, starting from the Visitors' Center at ground level on Interstate 29. The bus is small—not surprisingly; the road up the mountain is unpaved, single lane, with sharp turns and no guard rails—so only fifteen people a day can go, and the rule is first come, first served. But there it is.

HARVARD'S GREAT REFRACTOR isn't the only grand old instrument still around and available for visits. There's an even bigger refractor at the Yerkes Observatory near Lake Geneva, Wisconsin, a couple of hours' drive north of Chicago.

Yerkes, the man the observatory is named after, was not known for any particular interest in astronomy. However, he had made a fortune out of Chicago streetcars, and when someone suggested that building an observatory was a good way to perpetuate his name, he was quickly persuaded. (He also provided the original funding for the Yerkes Primate Research Center, near Atlanta, Georgia; the name of Yerkes is now well and truly perpetuated.)

The Yerkes Observatory may be the last big one that wasn't built on top of a mountain and about the last to be built around a lens rather than a mirror. It's a big lens—forty inches in diameter—and it is actually made of two pieces of glass of different chemical compositions, to minimize the chromatic aberration (translation: unwanted rainbow of colors) that lenses are prone to. For such a lens, four surfaces, two on each piece of glass, must be ground to close tolerance—another argument in favor of the mirror, which requires only one such surface.

Architecturally, the Yerkes Observatory was built in the majestic style of the late nineteenth century; as you approach, except for the sight of the great dome over one wing, it looks more than anything else like the main New York Public Library or Washington's Supreme Court building. Inside, the style is much the same—until you come to the great telescope itself.

At Garden Street, remember, the observer was carried along with the eyepiece of the telescope in a chair like a cherrypicker. At Yerkes they solve the problem in a different way. The whole floor of the observatory is on a giant, slow-moving elevator; you enter from the second floor of the building and then the floor rises, lifting you to the point where you can peer into the eyepiece.

The Yerkes telescope is by far the biggest I've ever been permitted to look through. The object in view when I was there was the globular cluster M–5, about 25,000 light-years from Earth. You can't see it with the naked eye. In the five-inch spotting scope attached to the big Yerkes instrument, it looked like a small, partly melted snowball. But then, through the forty-incher, it turned out to be a field of countless individual stars, and I don't think I've ever seen a prettier sight.

If you want to visit the Yerkes Observatory, they'll welcome you on Saturday mornings, when it is generally open to the public. You won't be looking through the forty-incher then, though, for at least two reasons: one, there's not much to see by daylight, and, two, generally when seeing is possible the Yerkes is still busy doing actual astronomy. But don't give up hope. Every once in a while there's some special occasion, and then your time may come.

WHEN ASTRONOMICAL OBSERVATORIES moved away from the cities of the East and Midwest, most of them wound up in California, with its (then) remarkably clear air and (still) decently tall mountains to perch telescopes on.

The first big California observatory was the Lick, on the peak of Mount Hamilton near San Jose. Like the Yerkes, its funding came from a wealthy man looking for something important to put his name on, James Lick. His first idea was to build a whopping huge pyramid right in the middle of San Francisco. Fortunately, he was talked out of that, and the Lick Observatory was the result.

The telescope James Lick gave the astronomers was a refractor, a whopping thirty-six inches across, but that was about the last large refractor anyone built. The next big California telescope was a sixty-inch mirror on Mount Wilson, in the Pasadena hills. (If you're flying in to Los Angeles International from the east you will likely see its dome as you descend.) That first Mount Wilson instrument saw first light in 1908, and worked so well that a one-hundred-incher was ordered almost at once, though World War I slowed down delivery and it didn't start service until 1919.

That hundred-incher has a grand history. It is the instrument Edwin Hubble looked through in 1924 when he made his discovery that some of those peculiarly fuzzy blobs that puzzled astronomers were actually galaxies like our own. And there's another interesting bit of history about Mount Wilson.

As you walk around the campus, you may notice some large ceramic pipe sections lying beside some of the paths. They may look as if they are meant for a drainage system, but they aren't. They've been there since 1931, when they were last used by Albert A. Michelson, as part of his experimental equipment (mile-long sealed tubes, with the interior air pumped out) to make the first accurate measurements of the speed of light. (For fun, ask whoever is guiding you around what the things are. He may have no idea—a good many of the astronomers don't—and so you can have the pleasure of explaining it.)

THE ONE-HUNDRED-INCHER worked so well that almost immediately astronomers began yearning for something bigger yet.

This time they thought *really* big. A consortium began drawing up plans for a titanic instrument with a mirror three hundred inches across. It was a daunting plan, and it would require dauntingly large amounts of money, but they persevered. They even succeeded in getting commitments to finance the thing. The year was 1928.

That was bad timing. The year 1928, alas, was quickly followed by 1929, the stock market crash, and the Great Depression. The financing collapsed. By the time it was possible to think of spending that kind of money on a telescope again, the astronomers had had second thoughts about the design, and when the great Pyrex blank for the new mirror was finally funded and cast, it had been cut back to a safer two-hundred-inch diameter.

Then World War II came along and brought additional delays; so it was 1948 before the two-hundred-inch Hale telescope on Mount Palomar—the first one to deserve its nickname as the Big Eye—received first light.

In some ways Mount Palomar is the ideal observatory to visit. For one thing, it's fairly convenient to drive to from almost anywhere in southern California; the observatory on Mount Palomar lies more or less midway between Los Angeles and San Diego, though well inland and up in the hills. To get there you drive through a few pleasant miles of orange and grapefruit groves and then climb the mountain. (Only during daylight, of course. If you came up the hill at night the scatter of illumination from your car headlights would thoroughly spoil some astronomer's day.) It's worth the trip just for the view, not to mention the pleasant pine woods nearby and the local wildlife. (Including the occasional mountain lion—but they almost never come near where the human beings are.) And, of course, the first thing you see is the immense dome of the Hale telescope.

Mount Palomar's Hale has done wonderful work for astronomers over the years—not least the first Palomar Observatory Sky Survey (referred to as POSS-I). That catalog of the heavens was sponsored by the National Geographic Society, and it became the basic road map for almost all astronomers for many years. (*Almost* all. Mount Palomar is located too far above the equator to be able to see all the southern skies, so there is a smallish area it could not cover.)

If you want to know how big the Big Eye is, one simple statistic says it all: In the decades when the Hale ruled as the world's largest and best telescope, it brought in more astronomical information than all the telescopes in the world had before it.

So when you look at it, view it with admiration. Don't expect to be allowed to look through the two-hundred-incher, however. You can't, and, as a matter of fact, no one else does, either, not even the working astronomers who operate it. The time is long past when people like the famous English astronomer Sir William Herschel and his devoted sister, Caroline, would take turns staying up until dawn, perched on an uncomfortable seat and freezing in the winter nights with their eyes glued to the telescope. The human eye is no longer part of the astronomical telescopic system. First the eye was replaced with camera film, capable of storing up photons over an hours-long exposure and thus producing better images than any man or woman could ever see; then spectroscopes were attached.

The spectroscope was a major revolution in itself. By breaking down the light from a star into its rainbow spectrum by means of a glass prism or a ruled diffraction grating, astronomers could observe the dark and light Fraunhofer lines that identify particular chemical elements and compounds. At once it was possible to determine just what stars were made of . . . requiring a quick revision of the old "Twinkle, twinkle, little star" nursery rhyme to:

Twinkle, twinkle, little star,
We can find out what you are
When unto the midnight sky
We the spectroscope apply.

The rush of technology didn't end there. By now telescopic "viewing" has become largely electronic, with charge-coupled devices that are far more sensitive than even the best photographic film. For that matter, the astronomers working a telescope like Mount Palomar's Hale may not even be physically present on the mountain while they are observing. Through the use of computer nets they may be many miles away during their observations, running the whole operation by remote control from their comfortable offices.

The telescopes are there, though, and they can be seen. Inside the Hale dome on Mount Palomar is a visitor gallery—it is one of those highly rewarding walk-right-in facilities—and from it you are looking directly down on the mighty Hale itself. Outside, all around its dome, you also can see such various specialized and somewhat smaller other telescopes, part of the same facility, as the forty-eight-inch Schmidt, also dating from 1948.

Why do the Palomar astronomers need more than one telescope? Because the telescopes do different things. The Schmidt, for instance, is a "wide-field" instrument. This means that it sees a larger expanse of the sky than the pinpoint that is visible at any one time with the two-hundred-incher. So telescopes like the Schmidt locate objects of interest, and then the two-hundred-inch zooms in to study them in detail.

Two of the nearby telescopes are worth a special look, because they represent the cutting edge of Earthbound astronomy. Neither of them is very large in itself, but they work as a team. The images from the two are hooked together, so that astronomical objects can be studied by means of interference patterns.

How do you do that? With great difficulty. It has been done for some time with radio telescopes (as we will see soon), but to do it with visible light is a task orders of magnitude harder.

But it's worth the effort. Linking two optical telescopes in that way permits studying details so fine that normally they could be observed only with an aperture equal to the distance between them—that is to say, with a mirror a hundred yards across!

I don't mean that you can get two-dimensional *pictures* of that sort. The *amount* of light received by the two telescopes isn't increased. But by studying the interference patterns thus obtained, astronomers can make measurements with exactly that sort of precision. I'll talk a little bit about what that means when we come to Hawaii's great new Keck telescopes.

THERE'S SOMETHING ELSE you can see from the top of Mount Palomar that is a lot less uplifting. That is the strip of gas stations and Taco

Bells along the highway, all of them bright with neon signs that produce unwanted light that strays into the observatory's instruments.

So even Mount Palomar isn't remote enough for the best seeing anymore. The light pollution doesn't come just from the highway clutter; even worse is the skyglow from the streetlamps of the city of San Diego, miles away.

For a time San Diego's civic authorities were obliging enough to replace some of their street lighting with low-pressure sodium lights as a favor to the Palomar astronomers. Those streetlights were just as bright, but happily for science, they emitted light only in one narrow band of the spectrum. That meant that astronomers could filter the resulting glow out of their observations. Not all San Diegans were pleased with that arrangement, though; they found the yellowy color of the low-pressure sodium light unflattering, and so a number of residents continue to press for a return to more conventional—and for the astronomers on Mount Palomar, far more crippling—systems. As of this writing the issue is still unresolved. I have friends in San Diego and I want nothing but good for them. All the same, I hope that wish is never granted.

LIGHT POLLUTION IS not the Earthbound astronomer's only enemy; the atmosphere itself is nearly as bad. The air we breathe is heavy, muggy stuff. Even when the sky isn't clouded over entirely, the air is full of water vapor and dust particles as well as all the obscuring pollution that we human beings throw into it.

One way to deal with that problem is to put our telescopes on the highest handy mountain. Then the murkiest of the air we breathe is well beneath. Of course, that raises problems of its own. (What doesn't?) One of the problems is that sometimes it makes enemies of such natural allies as scientists and conservationists, when their goals conflict. A battle of that kind is currently being fought in the Southwest, where for some years the University of Arizona has been trying to build observatories on Mount Graham, two miles up. What is wrong with that plan is that that particular mountain is the only remaining home of the endangered Mount Graham red squirrel. As of this writing, the battle between squirrels and stars has not yet been resolved.

There's a more general downside to putting a big telescope on top of a mountain. When you set about building your observatory, you have to deal with the logistical task of lugging huge, heavy, and delicate mirrors, as well as many tons of construction materials and equipment, up a mountainside that may or may not have anything like a real road. Then, even after the observatory is operational, the people who use it will have to make that same harrowing trip from time to time, if only to maintain their instruments.

That trip is not always comfortable. I can testify to this from personal experience, as a onetime visitor to the several large telescopes on top of Mauna Kea, on the Big Island of Hawaii.

THERE ARE NEARLY a a dozen great telescopes on Mauna Kea, and they are well elevated above most of the Earth's atmosphere. (Remember the Hawaiian solar eclipse I mentioned in the last chapter? The clouds we were able to dodge around on the *Independence* covered the entire Big Island, and there was no way to move that. So the tourists on the Big Island who had come to see the eclipse saw nothing at all. However, the astronomers on Mauna Kea, at 13,600 feet above sea level, were well above the clouds. Their only problem was a faint, high haze of particles from the recent eruption of Mount Pinatubo in the Philippines, but that didn't keep them from getting the most important of their observations—and they were delighted to do so, since their telescopes were not going to be in an eclipse track again for another several dozen years.)

Mauna Kea is a fine, big mountain. In fact, if you measure it from its actual base on the seafloor to its peak, it's probably the tallest mountain in the world. (Yes, the peak of Mount Everest reaches a higher point above sea level, but Everest has an unfair advantage. It starts higher. Everest's base is on the elevated Himalayan plateau, while Mauna Kea's base is at the bottom of the deep Pacific Ocean.)

So Mauna Kea is the site of choice for some of the world's greatest telescopes—including the monsters that are now the biggest of all, the wonderful twin Keck telescopes.

How big are the Kecks? Well, they've got the equivalent of four-hundred-inch mirrors, making their power of resolution four times as great as any other telescope in the world. They don't *have* four-hundred-inch mirrors, though. What they have is a next-generation offshoot of the Multiple Mirror design, a mosaic of thirty-six smaller mirrors joined together, functioning as if they were a single giant disk. It wasn't easy to make the Keck mirrors. For just one thing, each of those thirty-six segments had to be ground to a particular, unique curvature, one that no previous mirror had ever required, so that when they were assembled they would form one optically seamless perfect whole. But those problems were solved, and the Kecks are now a fact.

But that is only the beginning. There's a reason for building those two giants right next to each other: the hope that sooner or later they, plus several smaller instruments, will be electronically joined, like the two smallish ones on Mount Palomar. That's a complicated process, but if it works out as planned, the Kecks will then operate as one super-giant telescope, increasing their resolution by still another factor of ten.

That part of the project is still iffy. It's going to take more research and a few years of trial-and-error experimentation to make it happen, because nothing of quite that complexity has ever been tried before. But the odds are that it will succeed, and that means . . .

Ah, that means that, with any luck at all, the astronomers using that super-Keck system may be able to answer some of the most pressing questions of current cosmology. Is there an end to space? By resolving some of the most distant objects, and seeing if still more distant ones become visible, the Keck may tell us that. It will be able to study some of the strange, vast patterns that have emerged in the far reaches of the universe: the Great Wall in space, the aggregation of countless galaxies into a volume hundreds of millions of light-years long, 100 million light-years tall, 20 million light-years thick. Or the great void in Bootes, a volume of space 200 million light-years across, where few or no galaxies exist. The new Kecks may even answer, once and for all, the question of whether our universe will go on expanding forever or reach some limit, some time in the future, and then fall back into itself in the "Big Crunch."

Do such questions really matter?

Oh, maybe not, or at least not as they affect the course of our daily lives. Certainly we could go on cooking our meals and going to our jobs and watching our football games indefinitely without ever giving such things a moment's thought . . . but what kind of life is that?

CONSIDERED AS TOURIST attractions, the Kecks and the others atop Mauna Kea have the further advantage of being in that marvelous place to visit, Hawaii, with its fern forests and volcanos and waterfalls and perpetual balmy summer. But even the fraction of Mauna Kea that protrudes above sea level goes two and a half miles into the air, and it is no longer summer when you reach the peak.

The one and only time I ever visited the peak the Keck's had not yet been built, though there were a half dozen other big instruments already there. From that trip I learned a valuable lesson. I learned not to do it again.

I didn't go into the venture completely unaware. I had been told in advance that the road was pretty bad—but not quite *how* bad—so my wife and I rented a four-wheel-drive vehicle and took off. I was, I admit, a little deterred when I saw a big sign by the roadside—looking, more than anything else, like the one that warned Dorothy off the Emerald City of Oz. As I remember it, the sign said:

WARNING
Travel at your own risk.
No turnaround.
No water or food.
No toilets.
No repair facilities.
No communications.

The route up to the peak didn't really *look* all that discouraging. As far ahead as I could see the road was a broad, paved highway and the incline gentle. I plowed on.

That was just the beginning. A mile later the road had narrowed and begun to climb; a mile or two farther it had become an unpaved gravel track scratched out of the steep mountainside, with cliffs on one side and near-vertical five-hundred-foot drops on the other. There were no guard rails. From time to time there were boulders fallen into the road that needed to be steered around. Since the road was made of slippery crushed lava we skidded alarmingly. At two miles above sea level the car's engine began to stutter and gasp, complaining that it wasn't getting enough oxygen to breathe; when we entered the cloud layer the seeing got foggy, made worse by the fact that my Photo-Gray glasses, still responding to the bright Hawaiian sun, had darkened protectively and were taking their time about clearing again. I was squinting through a nearly impenetrable haze; and when at last we reached the top of the mountain, we emerged into a blizzard of high winds and stinging sleet.

I had been on enough mountaintops to know what they were like, so we hadn't failed to take some precautions. We'd brought sweaters along. But we hadn't brought anoraks, mittens, boots, and wooly hats; and we were *cold*.

We didn't go into any of the domes. We didn't stay very long at all, in fact. We snapped a few pictures, turned around, drove gingerly back down the mountain; and an hour later we were gently sweating as we soaked up some rays by the pool of our hotel on Banyan Drive.

It could have been worse, though. In fact, it was worse the next day for a party of French astronomers; the storm intensified, sections of the road washed out, and the astronomers were stuck on top of the mountain for five days, surviving on cold water and Hershey bars.

I still think the trip was worth it. I even think that, in spite of what I've said, next time I'm on the Big Island I'm going to see if I can get up there again, because I really do want to get a look at those wonderful Kecks. The road to the top is a little less awful now. At least, the last mile or so has now been paved—not so much for the sake of the traveler but to cut down the amount of dust that the car traffic was throwing into the air above the telescopes.

All the same, next time I'll leave the driving to someone else.

THERE ARE PLENTY of other Big Eyes in the world. The Russians have one in the Caucasus Mountains that for some years was the world's biggest, in fact—a mirror 240 inches in diameter, exceeding Palomar in size (but not in quality, because its optics are pretty poor). There's a new one in the Davis Mountains of Texas, the William Hobby–Robert Eberly 440-inch giant, out-Kecking even the Kecks with its ninety-one hexagonal mirrors and what they claim to be the darkest skies in America. (And, yes, with a visitors' gallery open to the public.) And astronomy is a growth industry in Chile, where there is a cluster of giant telescopes atop the mountains around La Silla, not to mention some fine Australian instruments.

But I can't tell you much about any of these from personal experience, because they're not all that convenient for the casual tourist to drop in on and I've never had the luck to visit them, and anyway there are some other kinds of telescopes entirely that are a lot more tourist-friendly.

For instance, there are the radio telescopes of the world.

A RADIO TELESCOPE doesn't look at all like any of the optical Big Eyes. It doesn't have to be on a mountaintop, either. Radio astronomers are lucky in that the kind of photons they "see" by come unobstructedly right down to the surface of the Earth—even on cloudy nights; even, for that matter, in daylight.

Radio telescopes look so unastronomical that for a good many years I frequently drove by one of the most famous of them without even knowing what it was. That one was a relatively small instrument that sat by the side of a road a mile or two from my then home in Middletown, New Jersey. It looked mostly like a giant, malformed tuba.

That radio horn belonged to Bell Labs. In 1964 it was being used by a couple of Bell scientists, Robert W. Wilson and Arnold A. Penzias, for the purpose of investigating radio interference in the ionosphere, the uppermost layer of Earth's blanket of air. Their researches weren't going well. It seemed that there was something wrong with their horn. No matter what they did, Wilson and Penzias kept getting a steady buzz of interference that they couldn't account for. The noise did not come from any particular part of the sky, which might have been explained by a radio object like a bright star. Their best bet, they thought, was something wrong with the instrument itself.

So Wilson and Penzias did their best to make sure the instrument was working right. They checked the electronics. The horn sat exposed to the elements—and to wandering birds—so they spent one long day cleaning pigeon excrement out of it. Nothing helped. The buzz remained.

Then one day, more or less by chance, they happened to mention their problem over the phone to some colleagues at Princeton University, Robert Dicke and Philip Peebles—who exploded with excitement, for their own current investigations had to do with the Big Bang that lay at the beginning of our universe, and they had just decided, on theoretical grounds, that there should be some detectable residual radiation from that colossal event. Theory predicted that over the billions of years since the universe began, the unimaginably high temperatures of the Big Bang should have cooled to an average temperature throughout the universe somewhere not far above absolute zero. That residual heat should be detectable in microwave radiation, Dicke and Peebles thought. They had been planning to build a radio horn of their own to look for it, since it should produce just such a constant hum.

Well, it did. Wilson and Penzias had indeed detected that remaining echo of the Big Bang, and it led to a Nobel Prize for both of them.

Since then other radio astronomers have repeated their observations, with increasingly fine instruments. The best of them have measured that cooling temperature much more accurately—it is pretty well established now that the average temperature of the universe is

2.726 Kelvin, as scientists put it. (Which is to say that it is 2.726 degrees on the Celsius—or what we used to call the Centigrade—scale above absolute zero. And absolute zero, the temperature that is so cold that nothing can ever get colder, is 273 degrees below the Celsius zero we're used to, the one that measures the freezing point of water. If you're not in the habit of using the Celsius scale, absolute zero is also 459 degrees below zero Fahrenheit. Either way, it's *cold.*)

This residual radiation is one of the best proofs yet found that the Big Bang theory of the early history of the universe is right—that is, that the universe popped into existence from nothingness 14 or so billion years ago, give or take a few billion, and has been expanding ever since. (Please don't ask *how* that happened. Scientists have theories about it, pretty hairy ones, too, but nobody really knows for sure.)

Actually none of the scientists mentioned so far was the first to suspect that that Big Bang residual radiation should still be around. That honor belonged to George Gamow, who had proposed it as early as 1948; but unfortunately Gamow got his arithmetic wrong, estimating that the temperature should be around 50 Kelvin instead of less than 3.

So the actual discovery was made by Wilson and Penzias, pretty much by accident. Which proves once again that scientists can find wondrous things, even when they aren't looking for them—provided only that they keep on looking everywhere for everything.

MY PERSONAL FAVORITE among radio telescopes is a lot more spectacular to look at than that Bell Labs horn. In fact, it is the biggest radio telescope in the world. Indeed, it is by a long shot the biggest telescope of any kind, and it is located in Arecibo, Puerto Rico.

As with the optical instruments on Mauna Kea—and so many other great telescopes—the Arecibo telescope is sited in a particularly pretty part of the world, up in Puerto Rico's mountains. The radio astronomers didn't choose the site because of its beauty, though. They chose it because geology had done a lot of their construction work for them. As those Puerto Rican mountains were being born, in great volcanic eruptions many millions of years ago, the molten rock that formed them trapped vast quantities of gases. When the rock cooled the gases remained, forming huge bubbles of rock under the surface. Some of those bubbles are still there as huge underground caves. But as erosion wore away the surface of the mountains, some of the bubbles were exposed as large, hemispherical valleys—almost exactly the right shape to form a radio dish.

So all the astronomers had to do was pick out a particularly shapely valley, line it with a metal grid to reflect the radio waves to a point, and hang a receiver overhead to detect them.

Well, it was a little trickier than that, to be sure. The valley surface was not a perfect curve, and so it had to be bulldozed to the correct curvature. The metal grid, in its present configuration, made up of forty thousand perforated aluminum panels, had to be laid carefully, and as tropical vegetation does its best to grow up through the panels, it has to be mowed from time to time. That overhead receiver is no slouch, either. It weighs six hundred tons and has to be suspended from cables moored to the rim of the valley. (There's a catwalk along one of the cables to permit access to the receiver. I walked out on that catwalk once, when no one was looking. Being there was a lot like looking down from the top of a skyscraper, except that you didn't have the reassuring bulk of the skyscraper between you and total destruction at the bottom of the valley.)

Such a serendipitously preformed telescope mirror has one serious defect. One of the qualities you would like to have in any telescope is the ability to point it at whatever you wish to look at. You can't do that with mountain valleys, though; they are not steerable.

That problem can be dealt with. By electronic manipulation, the suspended receiver can "move" the telescope's field of view to some extent north and south in the sky. The east-west problem is taken care of by the rotation of the Earth. Over every twenty-four-hour period, the fixed dish at Arecibo automatically scans the whole circle of the heavens.

It's a marvelous instrument. How sensitive is it? Well, if there happened to be radio astronomers somewhere in the M31 Galaxy in Andromeda, and if they had an instrument as powerful as the one at Arecibo, the two instruments could exchange radio messages. (After all, a radio transmitter antenna is only a receiver antenna in reverse.) To be sure, the distances between them would make it a very slow dialogue, since at the speed of light it would take the radio waves two million years to go each way.

As a matter of fact, in the fall of 1974 the Arecibo astronomers did try their hands at something like that sort of long-distance telephone call, though on a considerably smaller scale. Their experiment was part of the early SETI—Search for Extraterrestrial Intelligence—

project. They transmitted a three-minute burst of binary numbers to the star cluster M13 in the constellation of Hercules.

Why did they choose a star cluster? Because they are densely packed groups of many thousands of stars, giving many possible targets for the price of one. Why pick M13 in particular? Because M13 happens to lie conveniently within the belt of sky that is in easy range of Arecibo's fixed upward stare. Why transmit binary digits? Because any alien radio astronomers who happened to receive them would know at once that only intelligence could produce such a signal. Presumably as soon as they heard it they would fire back some sort of reply.

However, "at once" is still not very fast. Although M13 is not particularly distant as astronomical objects go—it is even faintly visible to the naked eye—it still lies about 34,000 light-years away. That three-minute burst of binary numbers that was transmitted from Arecibo—think of the message as a skinny line of radio photons 3 × 60 × 186,000 miles long—is still chugging away between here and M13. At the speed of light it will not reach the cluster until around A.D. 35,974. And the earliest date when we could hope for a reply would be around the year 69,974.

Of course, the Arecibo astronomers knew that. They aren't hanging over the phone, waiting for it to ring. They were simply trying out the procedure to see how it would work.

Alas, astronomers are less likely to perform such experiments now than they were in 1974, for SETI has fallen on harder times. What with the pressures of congressional budget-cutting, SETI's relatively tiny appropriation was an easy target for termination . . . and every fathead in Congress (especially the ones who were most diligent in voting pork-barrel projects for their own districts) had a high old time doing comedy routines about looking for "little green men."

All is not lost. Private contributions have made up most of the slack. Even Arecibo has a dedicated SETI receiver on the platform that hangs above the great radio dish, paid for with private funds. And (as we will see in a moment) the Search for Extraterrestrial Intelligence still continues.

AS TELESCOPES GO, Arecibo's is definitely the one with the largest aperture in human history. Its dimensions are not likely to be exceeded on the Earth's surface at any time in the future, either, if only because that's no longer necessary. Radio astronomers have learned a new trick.

Instead of building ever-wider dishes, they have devised electronic ways of linking any number of smaller radio dishes into a single system, somewhat in the way that the future optical systems of the Kecks will be organized. In that way the instrument's resolving power is increased in proportion to the distance between the farthest members.

One of those instruments is in Socorro, New Mexico, an hour's drive south of Albuquerque. It's called the Very Large Array, and it contains more than a dozen individual dishes, each mounted on its own little railroad car so they all can be moved about for focusing. (If you saw the movie *Contact* you saw the Very Large Array; it is where Jodie Foster sits brooding when without warning the first message from space comes in.) It does have one problem. Railroad cars need tracks to run on; and tracks need ties to hold them in place; in order to save money, the VLA's builders picked up a bargain in secondhand wooden ties from abandoned rail lines.

Cheap is cheap. The old ties are beginning to wear out. Some of the antenna cars are already marooned where they stand, reducing the VLA's flexibility. Sooner or later the whole layout of tracks probably will need to be ripped out and relaid, or the VLA's components will be frozen in place. No, it never pays to cut too many corners.

HARVARD'S GARDEN STREET installation was the first real observatory I had ever seen; the second was owned and operated by the same people. When Cambridge's increased light and pollution wrecked the seeing at Garden Street, they set up a new observatory out in the boonies. That was the Agassiz Station, now known as the Oak Ridge Observatory, in the little town of Harvard, Massachusetts. By current standards Oak Ridge is a fairly modest enterprise. When I visited it in the mid-1960s it had one nice-size optical reflector and, right next to it, an eighty-four-foot steerable radio-telescope dish.

Before I got to Harvard, I had known that radio telescopes existed, but I didn't know much about them. It turns out that you don't look through a radio telescope. You don't even take pictures with it. I hadn't exactly known that, and was slightly disappointed, as well as bewildered, to see what the data a radio telescope produces looked like. It was nothing but strings of numbers, nothing I could recognize as a picture. The numbers represent the frequency and intensity of the radio emanations from the object being studied. They contain all the information about the subject, but they sure aren't anything to look at. It takes a lot of computing to transform them into something the eye can recognize.

Now, with the much more powerful computers that are available, that transformation is done routinely and quickly, color-enhanced as desired, and some of the prettiest astronomical pictures around are from radio astronomy. But remember, my first look at their output was in the early 1960s, and the computers of the time weren't up to transforming tables into pictures.

Even by the time of my visit Oak Ridge had fallen well behind the cutting edge of observational astronomy. Both its optical and radio telescopes were tiny compared to giants elsewhere in the world. They still did good work, but they were definitely well behind the leaders in the field.

But then—on the day before Halloween in 1995—Oak Ridge became the cutting edge again, in that special subfield of astronomy, the Search for Extra-terrestrial Intelligence.

PEOPLE BEGAN SEARCHING for ETI long ago. How long ago? Probably about the time human beings first realized that those funnily moving objects in the sky they called planets were actually worlds themselves. Since our world has people, mightn't those others, as well?

The first serious searches—well, *fairly* serious searches—didn't employ radio, because there wasn't any. They started when telescopes got powerful enough to make out some surface features on Mars. To Percival Lowell in the nineteenth century, the lines he thought he saw on the surface of Mars seemed much too straight and regular to be natural. He was convinced they had to be canals, probably meant to irrigate Martian farms with water from the planet's polar ice caps. Canals had to be dug by somebody, and anybody smart enough to do that kind of huge-scale engineering was probably enough like our human selves, he reasoned, to want to strike up a conversation.

Any number of schemes were proposed to signal the Martians that we wanted to talk. One idea was to dig out vast, geometrically arranged trenches in the Sahara, fill them with petroleum, and set them alight some night when Mars was in the sky. Another was to turn the lights on and off in a few European cities, preferably three of them arranged like the vertices in an equilateral triangle. None of these ventures ever came to fruition, because about then radio came along, and people began listening.

Probably the first to hear anything he hadn't expected was Karl

Jansky, also at the Bell Labs, who was equipped with a primitive little antenna that he set up not far from where the Wilson-Penzias tuba now stands. Like many a scientist since, Jansky thought for one heart-stopping moment that he had indeed made contact—only to realize that what he discovered was the equally important fact that many objects in space radiate in the radio wavelengths. Many other people with a sensitive radio at hand wistfully tried their luck in the years that followed—again with occasional startling moments that generally turned out to be something like a Texas taxi dispatcher. And not long after World War II a few astronomers—Frank Drake the leader among them—began a systematic program of listening purposefully on selected wavelengths.

Things got a little easier when computers got more powerful, but then they got much worse again when the bulk of the money, which had come from the federal government, was abruptly cut off.

Enter the Planetary Society, a nonprofit organization with a membership willing to kick in what they could afford. Aided by a handsome grant from Steven Spielberg, SETI at last could afford a state-of-the-art computer. Then, in October 1995, the Oak Ridge Observatory offered to dedicate its old radio dish for the purpose, and right now, day and night, it is running the program they call BETA (Billion-Channel Extraterrestrial Assay), sampling 250 channels per cycle.

So everything's now fine, right? The SETI program is bustling right along, and they don't need any more help?

Wrong. They do need help. They need volunteers very badly. Interestingly enough, one of those volunteers could very well be you.

IT'S HARD TO convey just how tough SETI's task is. There are hundreds of billions of stars in our own galaxy (not to mention the 100 billion or so other galaxies external to our own.) Some of them, we know now, do have planets. Surely *some* of those planets *somewhere* ought to have life enough like our own to be worth talking to. And presumably some of them will have developed radio.

But looking for their radio messages is like hunting for a needle—no, for the *eye* of a needle—in a haystack the size of Earth. The universe is a very radio-noisy place. Just about every distant object in the sky is throwing out its own radio emissions, and how do you tell which is some distant entity transmitting a message and which is just a patch of heated gas?

Although the SETI people now have massive computation resources, they are not without limit. So, as we have seen, the astronomers have made some simplifying assumptions about the most likely places where an extraterrestrial radio message can be found.

But what if the assumptions are wrong?

In order to check the *un*likely places, what SETI needs is more computer time. A lot of it. So a University of California–Berkeley astronomer named Dan Wertheimer has launched a program he calls SETI@home. He's looking for volunteers with desktop or laptop computers—so far just over 1 million of them have signed up—to help check the data in the places where the official searches don't look.

If you offer Wertheimer your help (at setiathome.ssl.berkeley.edu), he will send you a program and a batch of data. Then all you have to do is load it into your computer and relax. It won't interfere with your own use of your computer. The program will run on your machine only when you aren't using it, and it will do it automatically like a screen saver. Then, when your batch of data is processed, you send the results back to Wertheimer and get another batch. Then SETI's computers will analyze what you have sent them to see if you happen to be the one who finally has detected the first communication from an alien intelligence.

Oh, it's an extreme long shot, of course. But SETI@home is about as painless as any research opportunity for a nonprofessional can ever be . . . and think of the possible reward!

RADIO WAVES AND optical light are not the only messengers from the skies that astronomers would like to study. There is no part of the electromagnetic spectrum that does not convey information—radio, visible light, infrared, ultraviolet, gamma rays, X rays, and all.

Indeed, some sources of information aren't "waves" at all (not counting the "quantum reality" rule that says that, in the final analysis, *everything* is both particle and wave . . . but we don't want to go into that now, do we?). To make use of these non-electromagnetic sources, there are such bizarre instruments as the world's handful of neutrino "telescopes."

Bizarre? Well, what else would you call a telescope that functions best when it's buried deep underground?

What makes these neutrino telescopes bizarre is the nature of the very particles they "see" by: neutrinos. Neutrinos, which are produced by intense nuclear reactions, such as those that occur in the heart of stars, are tiny, chargeless, and all but massless particles. Because they are so small and electrically "neutral," they make their presence known only very rarely by interacting with matter. There are a lot of them around, though: something like 30 *million* of the things in every cubic foot of space throughout the universe. According to some theories, lightweight as they are, the total bulk of all the universe's neutrinos may far outweigh that of all the visible stars, gas clouds, and planets, but they are not at all easy to detect. Every second billions of them sleet through your body, and you never notice that it's happening. Of the vast number of them that strike the Earth, nearly all go right on through—through atmosphere, surface, mantle, and core—and come out on the other side. When they emerge they have left no trace, and, not even slowed down by the experience, they keep right on going.

No one has ever seen a neutrino or even detected it directly. But once in a very great while some single neutrino, while usually passing through the component parts of any atom as if they were not there, happens to strike one of its subatomic particles head-on. Then, like the colliding particles in one of the giant accelerator rings in the laboratories I spoke of earlier, it produces a shower of other particles. It is those other particles that can be detected, just as is done at Fermilab or Brookhaven.

Such collisions do not, however, happen very often. To make things worse, all sorts of other particles—cosmic rays, for instance—are striking the same target and chipping off their own showers of by-products. Anywhere on the surface of the Earth there are many more of those other events than there are of neutrino collisions. It is all but impossible to find the events you're looking for among the constant shower of unwanted others; so if you want to observe neutrinos, you not only don't mind the filtering effect of Earth's atmosphere, you want as much heavy-duty filtering as you can get. The only sensible place to construct a neutrino telescope is underneath some large mass of matter.

What kind of matter? Well, you have two choices there. Water will do if you go deep enough. For that reason, neutrino astronomers are now experimenting with dangling strings of detectors in Lake Baikal in Russia, in the Aegean Sea near Greece, and in the Pacific Ocean off the coast of Hawaii. There is also a similar experiment—called AMANDA, for "the Antarctic Muon and Neutrino Detector Array—a mile and a quarter down in the Antarctic ice, which is of course nothing but water in its frozen state.

Neither water nor earth filters out all the incoming particles from space that may be producing false results, but there's a trick that helps out here. When the odd neutrino does hit a subatomic particle in the water-tank detector, it produces a tiny, tiny cone-shaped flash of blue light, called Cherenkov radiation. If the water is transparent enough, the neutrino detectors, which are really nothing much more than tiny cameras, will record the flash and determine which way the cone points.

That's the trick, because from that information you can tell which direction the impacting particle came from. If from above, it may have been just one more confounded cosmic ray; but if it came from below, it had to go clear through Earth to get there and thus can be nothing but a neutrino.

The alternative to working underwater is to build your neutrino telescope underground, the deeper the better, where more sophisticated intrumentation is possible and can give a lot more information.

Abandoned mines are a favorite place for neutrino astronomers to build their instruments. The first major neutrino telescope was down deep in the old Homestake gold mine in South Dakota. The mine owners were kind enough to be cooperative, provided the neutrino astronomers didn't get in the miners' way—for which you can't really blame them, since Homestake produces about $100 million worth of gold a year. Or you can hide it under a mountain, which is where the biggest European instruments are: under Mount Andyrchi in the Russian Caucasus (that one a Russian–American joint project) or at Gran Sasso in the Italian Appenines.

The Gran Sasso operation was a target of opportunity, quickly seized when Italian astronomers learned that highway authorities were digging a tunnel through the mountain. The astronomers knew a good chance when they saw one, so they persuaded the government to let them dig an additional side cavern off the middle of the tunnel for their observatory—nice and deep, and made all the more convenient by the fact that they could drive right to work through the tunnel.

Of course, that produced its own problems. The rock of the Appenines is granite, which has an undesirable trait. Granite emits its own radiation (from tiny quantities of radioactive elements trapped inside it as it forms), and even that tiny radiation could confound the astronomers' observations.

The solution to that problem was to line their observation chamber with thick-enough lead to keep all that stuff out; but the problem that arose from *that* (you see, it never stops) was that if lead lies around on the surface of the Earth for any length of time, it picks up secondary radiation of its own, from cosmic rays, or nuclear bomb testing, or such occasional events as the Chernobyl accident.

There, too, the Gran Sasso neutrino astronomers were lucky. It happened that marine archeologists had just discovered a freight vessel from the first century B.C. at the bottom of the Mediterranean. Its cargo had been lead ingots, from the ancient Sardinian mines, and those 1,900-year-old ingots had lain underwater long enough to lose most of their cosmic-ray–induced radioactivity (and, of course, they had never been exposed to man-made sources). So the Gran Sasso scientists commandeered the cargo and swiftly fabricated it into shields for their instrument.

(Steel also picks up induced radiation, although it is not quite as serious as in lead. Still, future neutrino astronomers are eyeing a source of radiation-free steel that currently lies at the bottom of Scotland's Scapa Flow, where the Germans scuttled the battleships of their High Seas Fleet after World War I.)

The biggest and newest neutrino telescope around, still under construction at this writing, is in a copper mine in Sudbury, Ontario, a couple of hours' drive northwest of Toronto. The town fathers of Sudbury are hoping for a great deal from their new neutrino center. They feel they need a break. In environmental circles Sudbury had a somewhat tarnished reputation for a good many years; the exhaust stack from its copper smelters was notorious as the dirtiest stack in the world, all by itself producing 1 percent of all global air pollution. Its toxic fallout damaged forests and farmlands for miles around and even helped kill lakes in New England from acid rain. Now that horrendous stack is stilled, but so is much of the copper business. The town is feeling the effects. So Sudbury is hoping tourists will come to see the neutrino center, and to that end they are planning a science museum and a lot of other attractions.

Which will be worth having when people like us come by to visit, because, to tell the truth, a neutrino telescope is not in itself the most interesting spectacle one could hope to see. The "telescope" itself is nothing more than a large tank of some liquid—at Homestake it was ordinary dry-cleaning fluid; at SAGE in the Caucasus, liquid gallium—sometimes chemically analyzed to see what changes have taken place, generally surrounded by photocells to catch each flash of light as the occasional neutrino strikes an atomic particle in the tank of fluid. The flashes are far too quick and tiny for the human eye, and there aren't very many of them. To make it worse, even the best of the current neutrino instruments can detect only one of the many varieties of neutrino, the so-called electron neutrino.

But when the astronomers do see something, think what they are seeing! Starlight—any kind of starlight, in any frequency—comes from the surface of a star. The neutrinos come from a star's very heart, and they pour out in the greatest quantity at the most exciting times in the star's history, as when its heart is exploding as a supernova. Since neutrinos pass so readily through almost any obstacle, some of them come to Earth from very great distances. When Supernova 1987A flashed into existence in the Small Magellanic Cloud, neutrino observers in several parts of the world recorded pulses from that giant blast.

NEUTRINOS, RADIO WAVES, and optical light can be detected by planet-bound instruments, but there are many kinds of radiation that cannot. Unfortunately for astronomers on Earth, making use of these other sources is not easy. In fact, until quite recently it was wholly impossible, since our planet's atmosphere is opaque to some of the most interesting frequencies. (Unfortunately for the astronomers, but a real break for the rest of us. Those higher-powered frequencies, from extreme ultraviolet on, are uniformly destructive to living things, and without the opacity of the atmosphere to shield us from that deadly radiation we would never have been born.)

So most of the very newest observatories have fled to points even farther than mountaintops. They are orbiting out in space itself. There instruments like the Compton Gamma-Ray Observatory and a number of X-ray and extreme ultraviolet instruments are studying the heavens in the forbidden frequencies, while even optical instruments like the Hubble are getting previously unobtainable images of tiny and faint objects.

For seeing objects in our own solar system, there is an additional trick.

The most useful optical telescopes that are studying the rest of our solar system now aren't anywhere near as big as many on Earth, but they see better because rockets bring them closer to their targets; they are mounted on spacecraft.

So if you want a really good close-up look at, for instance, the moons of Jupiter or the rings of Saturn, the best place to see them as the pictures come in is at the Jet Propulsion Laboratory in Pasadena, California, where the pictures are received as the spacecraft transmit them back to the waiting astronomers on Earth.

But that brings us to the subject of the next chapter. It is all very well to look into space, but it is even more exciting to go Out There . . . and so let's go on to talk about things to see and do in connection with the world's space programs.

Chapter 4
REACHING FOR THE STARS

The Space Program

The heart of the American space program is the Kennedy Space Center at Cape Canaveral, Florida. If you haven't happened to visit it in person, you've surely seen it on television, which has faithfully reported on its triumphs—and sometimes its great tragedies.

My own most memorable visit to the Cape had aspects of both. It was a night in the fall of 1972, and I wasn't exactly *on* the Cape. I was standing on the deck of the Holland-American cruise ship *Rotterdam*, as it lay anchored a mile or so offshore, and I wasn't alone. Among the others crowding against the rail with me that night were Isaac Asimov and Carl Sagan, Robert A. Heinlein and Hugh Downs, Theodore Sturgeon and Norman Mailer, and forty or fifty others who had been invited on that cruise. Every one of us kept our eyes fixed nervously toward the west, where we could see the distant lights of the gantry of the giant rocket that was about to take off. We couldn't hear the countdown, and we were much too far away to be able to make out any of the thousands of human beings who were on the shore, most of them herded into the grandstands a couple of miles from the launchpad and waiting, just as we were waiting, for what was about to happen.

Then, without warning, it happened. We saw a sudden burst of a different kind of light, ruddier but brighter, suddenly flaring at the base of the giant Saturn V rocket.

As our hearts began to pound, the rocket began to rise. It climbed past the top of the gantry, picking up speed. It wasn't until then that the deep basso-profundo thunder of the rockets rolled over us, even at that distance seeming to shake the fillings in our teeth. As the rocket rose up over the Cape it looked like the Sun coming up—but doing it in the west!—and climbing up the sky faster than any sunrise. The light of the rocket flare was now so bright that we squinted to look at it. I turned for a moment to glance at the others, and swore at myself for not having thought to bring a camera, so that I could have taken a picture of those faces illuminated by that spectacular light. When I looked back the rocket was climbing up over our heads, then out over the Atlantic, the point of light dwindling.

Then it was gone. The men aboard it were on their way to the Moon.

That was the triumphant part. The tragic part was yet to come.

I had known before I boarded the *Rotterdam* that this would be the first nighttime launch, and thus visually even more exciting than any of the ones had gone before. What I hadn't known in advance was that this *Apollo 17* launch would be the last there was. Congress had lost interest in Moon landings. The launches that had been planned to follow were canceled. The couple of Saturn V rockets and Apollo spacecraft that were left over stood unused, and before long even the assembly lines that had built them were abandoned.

That was the end of the Age of Men on the Moon. It had begun with the touchdown of *Apollo 11*'s Eagle lunar landing module on Mare Tranquilitatus on July 20, 1969, and altogether it had lasted just a little more than three years.

THE CAPE IS still there, of course, and it's still one of my personal favorite places to see science in action.

I've been there for other launches—for the first test launch of the immense Saturn V rocket that carried dozens of astronauts into orbit and for a couple of shuttle launches more recently—but it doesn't require the spectacle of a launch to make the Cape an exciting place to visit. It has plenty to see at any time. What's more, it's a great destination for a family vacation in the sun; it's an easy drive from Orlando, with Walt Disney World and all the other theme parks and entertainments there, and it welcomes visitors.

The first time I went to the Cape was in 1964. No launch was scheduled. I just wanted to see what the place was like. Everything was new—even the name of the Kennedy Space Center was new, because the president who had conceived the whole project had been murdered in Dallas only months before. Even as early as that I had already seen enough films and TV programs to recognize the gantries, the great multiwheeled flatbed truck that inches the rockets from the immense Vehicle Assembly Building to their launchpads, the VAB itself. The real astonishment came with perceiving the *size* of everything, starting with the VAB, the place where the Saturn Vs were painstakingly put together (and where, later on, they were mated to their Apollo capsules). The Vehicle Assembly Building is the largest enclosed space ever built by man, so huge that wisps of cloud form in its upper reaches, giving it its own internal weather. The spaceport stretched for miles upon miles between the Atlantic Ocean and the Banana River. The grandstands that held the invited guests at a major launch were themselves more than two miles from the launchpads—and not an inch too far. When I returned to sit in those stands a little later for that first Saturn V launch, that deafening white-noise roar of the rockets shattered an electric lightbulb over my head. Scattered across the expanse of the spaceport were all the ancillary buildings, the places where they worked on the parts of the spacecraft, the structures where they made and liquefied the hydrogen and oxygen that would burn to produce the rockets' thrust.

Even at that early date, the Kennedy Space Center was making itself as visitor-friendly as possible. In one of the buildings there was an auditorium where they showed films of the launches of the past—and some nonlaunches as well. For a while in the 1950s the National Aeronautics and Space Administration had experimented with exotic fuels for their future spacecraft. That idea didn't work out very well, and the films show why. It is a comical (if saddening) experience to view those films of one spacecraft after another rising a few yards and then falling back, or toppling over on the pad, or simply never moving at all except to blow up, with great plumes of exploding vapors in all the colors of the rainbow from the odd chemicals that were being burned.

That was then. Now it's even better for the visitor—more access, more displays. There's the 100,000-square-foot Apollo–Saturn V Center containing an actual Saturn V—longer than a football field, weighing as much as three 747s—and reproductions of the original firing room and of the lunar surface for simulated space flight and lunar landing. There are the back-to-back IMAX theaters, and the Rocket Garden with its examples of all the major vehicles, and the Space Station Processing Facility, where the parts of Space Station Freedom are brought from wherever in the world they're made and processed to be ready for lifting by shuttle into orbit. This is a working facility, not a museum, but visitors can watch the work of preparation as it goes on and then retire for a cup of coffee in the building's own cafeteria.

All this is open to the public, year-round. If you want information about tours you can phone or write—they get a lot of mail there, so they have their own zip code; the address is simply Kennedy Space Center, FL 32899.

YOU MAY EVEN be able to watch an actual launch, but for that you must request a car permit three months in advance. You will be assigned a parking space at a suitable distance from the launchpad, and there will be bleachers nearby where you can sit and watch the happenings. However, take a few precautions. If it's a daytime launch, take a hat and sunblock. If nighttime, bring all the mosquito repellent you can carry. It won't be enough—the Canaveral mosquitos are numerous and thirsty—but then you'll get that wonderful nighttime pyrotechnic show. (And as an extra attraction, if you listen carefully after the sun goes down, you'll heard the booming of the Cape's population of alligators as they call for their mates in the nearby swamps.)

Even if you fail to get a launch permit, all is not lost. You really don't need to be on the site to get a fine view of the show. For major launches, when attendance is restricted, the far side of the Banana River is sometimes lined with the cars and campers of the uninvited. They aren't given a grandstand to sit on, but they're not much farther away from the action than the VIPs in the stands, and the view is just about as good from a camp chair on top of a van.

THE KENNEDY SPACE Center is where most American spacecraft are launched, but the Cape is only one dot on the map of American space installations. There are scores of others. If the spacecraft are manned, their crews will want to come home again sooner or later, for which they need landing places. Sometimes the spaceships may land at the Cape itself or at Patrick Air Force Base nearby; more often, though, they land on the wider, longer, more remote runways of the desert flats at Vandenberg Air Force Base in California. And every manned American flight is supervised and monitored by the Johnson Space Center in Houston, Texas.

That's where Mission Control is—in fact, it's where two Mission Control chambers are, the old one and the new. The old is the one that is familiar, to those who are old enough, from the Apollo flights, and it is preserved for its historic value. If no space flight is being controlled at the moment, visitors may be permitted a look in at the new, larger one, as well.

The whole Johnson Space Center has become a sort of space-program theme park. Visitors are not only allowed but actively catered to; beginning in the 1980s there has been significant remodeling and new building to make it all tourist-friendly. There are exhibits, a gift shop, an auditorium for ceremonies, lectures and films, an assortment of actual space vehicles to look at, and plenty of parking space for visitors' cars.

IN HUNTSVILLE, ALABAMA, it almost seems as though the whole town has been turned into a theme park. It was largely in Huntsville that much of the post–World War II early rocket design and development took place, before anyone but dreamers entertained notions of actually firing human beings out into orbit. Now it is a sort of open-air and frequently interactive space museum. Some of the best of the exhibits are like the thrill rides in an amusement park. There are counterbalanced teeter-totters in which you can experience some of the effects of weightlessness. For the opposite experience, the added weight of a takeoff thrust, there is a centrifuging installation like a merry-go-round, but a lot faster and more adventurous than anything you've ever done before. For that one you strap yourself into a sort of open vertical coffin at the end of a long beam. When it starts to revolve you take as many gees as you can stand—well, as many as they're willing to give you, with thought to their liability insurance. As you enter, the guides will tell you that it's better not to move your head abruptly when the gees are building up. (They give you good advice. Take it.) Outside, the grounds are punctuated with spacecraft of every kind, so that you can see for yourself just how astonishingly huge, for instance, a Saturn V booster rocket is. ("You mean *that* thing flies into *space*?") There is a museum of space curiosities that, when I was last there some time ago, included one living specimen of an astronaut among its exhibits. He wasn't a human being, though. He was the survivor of the two monkeys that had been America's first space-faring primates. (Alas, I am told that he has passed on since.)

SPEAKING PERSONALLY, I grew up with space travel in my blood, transfused there by all the science fiction stories I'd read. That meant that the kind of space travel I really cared about was the kind where somebody jumped into his rocket (or whatever) and headed for Mars, Ganymede, or some far star to visit it in person.

I wasn't alone in this. The American public, and American presidents, too, seemed to care fundamentally only about putting men (and ultimately women) in space.

Yet that's not really sensible. The basic reason for going into space isn't to drive a golf ball across the lunar surface or to practice aerobic exercises in free fall. There's really only one valuable commodity worth getting in space. That commodity is knowledge. *New* knowledge, the kind of answers to questions that you can't get in any other way, and don't even know what questions to ask because you don't know what you're going to find until you find it. And for that you don't really need to take human beings along.

Human beings are really a lot of trouble on a spaceship. They need to eat and drink and breathe, so you have to take large supplies of all those expendables along to keep them alive. They have the habit of excreting after all that ingesting, and that, too, means extra trouble. (Ask any astronaut what he thinks about microgravity toilets; ask any female astronaut how she liked the giant diapers that were for a time all NASA could devise to cope with female anatomy.) Human beings can't stand a great deal of acceleration, which limits how fast they can launch; they can't stand too much heat or cold, which imposes restraints on how their living quarters can be designed. And even if you spend all the effort and money and lift mass necessary to accommodate all those expensive and chancy needs, you then have to face the fact that really worthwhile space exploration may take years, decades, or longer—the two hardworking (but unmanned) Pioneer spacecraft have been out earning their pay for us for nearly thirty years now, and there are proposed missions that would take far longer. Most human beings really don't want to stay away from home that long . . . not to mention that prolonged confinement in a spacecraft has serious, possibly even fatal, consequences.

No, people aren't your best bet for space exploration. Remote-controlled or automated probes can do the same job cheaper, faster, and in the long run better.

(Oh, in the *very* long run, sure, we're going to want to have human beings living on the Moon, and on Mars, and in O'Neill–type space habitats and even manning ships that will venture out to the nearest stars. That's manifest destiny. But before we can start thinking seriously about any such long-range missions, we want to know an awful lot more than we do about how to get there and what the explorers will find when they arrive . . . and the best way to do that is to send robot probes out to do the exploring for us.)

So the real action in space comes from the swarms of unmanned satellites, hundreds of them in orbit at any one time, that we put up there to do *work*.

What kind of work? Well, there are weather satellites that give the best warning the human race has ever had of dangerous storms (and that made it possible for the captain of our cruise ship to thread his way through the Pacific cloud cover so we could view that wonderful solar eclipse). Communications satellites that put the world's television on the screens of six continents. Science satellites, like the Hubble and the dozen other telescopes, that scan the universe in other frequencies. The Global Positioning System, twenty-four satellites that make it possible for anyone in the world to invest a few hundred dollars in an instrument that, at any moment, lets the owner know just where she is on the Earth's surface. And, of course, the crowd of military ones: satellites for optical or radar surveillance, electronic-intelligence satellites that eavesdrop on hostile communications, early-warning satellites designed to spot missile launches, ocean-surveillance satellites to keep track of potentially enemy submarines, and many others. The American Department of Defense alone launches a satellite a month, and the rest of the world keeps pace.

Most of the satellites are invisible to even moderate-size telescopes, but, providing you know where to look, a fair number can be seen with the naked eye. The trick is in knowing where to look.

By international treaty, every nation is expected to inform the United Nations of the name, purpose, and orbital parameters of every launch. They don't always do it, though. Particularly in the case of military satellites, they may give false information, or give the right data for the launch itself but then change the orbit with additional burns—firings of their engines.

But, for some people, that just makes the hunt more interesting. Amateur sky-watchers in many countries make a hobby of trying to locate and identify these mystery spacecraft and exchange information about them.

The bigger the satellites are, of course, the easier they are to spot optically, even by naked eye. About the easiest to spot currently is the Russian Mir, big and relatively close. As it happens, I did see that one myself, while this book was being written, from a hotel room in San Francisco. I had spent the day at sessions of the American Association for the Advancement of Science, and while I was changing to go out to dinner one of the television weatherpersons announced that Mir would be rising in the southwest at 6:36 that evening. I was lucky enough to have a westward-facing window. Although clouds obscured part of the view, there it was, a tiny pale dot sliding upward through the cloud veil.

OF COURSE, MIR didn't come from the Kennedy Space Center. The Cape is America's premier spaceport, but it's not the only one in the world. There are others belonging to the Europeans, the Chinese, and the Russians.

It's fair to say that ours is the best. That's not because, or not just because, of any special American virtues. Partly it is simply that we were luckier than the others in our geography.

You generally want to launch your spaceships toward the east, because that lets you take advantage of the small but useful extra push you get from the eastward rotation of the Earth itself. (The exception to that is satellites that are destined for a north-to-south polar orbit. For them, that bonus of velocity from Earth's rotation isn't an asset, it is a liability that has to be gotten rid of. The launch track for them is better when it has a westward component; thus the American launches from Vandenberg in California.)

Another geographical consideration is that, for normal launches, the farther the spaceport is from the equator, the more fuel must be burned to lift the spacecraft into orbit. The maximum boost from the Earth's rotation is at the equator. It falls off drastically as you move toward either pole.

What's more, you would like to launch in such a way that there is a fair-sized area in the direction of launch that is not densely occupied. That's so that when the spacecraft drops its booster tanks or when there is an abort—as there surely will be from time to time, when the ground controller has to make the sad decision to press the destruct button and explode an errant spacecraft into scrap metal—those great chunks of metal will not rain down on some luckless city.

So the fact that the United States includes the state of Florida is a major break for us. The eastern coast of Florida is a nearly perfect place for sending spacecraft into orbit. It is far closer to the equator than any site in the other spacefaring countries, and what lies to the east of the Cape is nothing but the broad and largely empty Atlantic Ocean. When we launch from Kennedy Space Center any debris that results falls harmlessly into the sea.

The other major players are not so fortunate in any of these counts. The Chinese and the Russians do have a fair amount of land area at their disposal, and thus their spaceports, in northern Sichuan Province and in Tyuratam, respectively, are able to launch over sparsely populated areas. (But they, particularly the Russians, have done serious ecological damage to those down-range areas with a litter of scrap metal and toxic fuel residues.)

The Europeans don't have even that much to work with, since the continent of Europe doesn't possess any large sparsely populated areas. What the Europeans are forced to do is to ship the completed spacecraft somewhere else. At times in the past the French have launched from a site in the Sahara and the Italians from a converted oil rig in the Indian Ocean, just off the coast of Kenya. Those weren't good enough; so now the Europeans ship their completed spacecraft across the Atlantic to be launched from their own high-tech spaceport at Kourou on the coast of French Guiana in South America.

I mention these other spaceports just to keep the record straight, since I've never been to any of them. Nor, in all probability, will I, since none of them is a very convenient spot for the casual tourist to visit. French Guiana is remote; the Chinese spaceport is almost as closed to outsiders as their nuclear facilities at Lop Nor; the Russian launch facility, though no longer entirely forbidden to foreigners, is not easy to get to, and, by all accounts, when you do get there what you find is a dishearteningly rusting and decaying mess.

That's a real shame, because you have to give the Russians credit for what they accomplished in the history of space exploration. They did wonders with what they had. The Russians put the first satellite in orbit, and the first human being, too. A Russian spacecraft was the first to circle the Moon and send back pictures of its never-before-seen far side. The Russians explored Venus fairly successfully long before we did. (I'll say more about that shortly.) In general, what they accomplished with their space effort was truly amazing, considering the obstacles they had to face.

What obstacles were those? The geographical ones were bad enough, but they also had serious problems with politics and military security. I don't mean anything as trivial as the American variety of red tape, cumbersome as that sometimes was. That was as nothing compared to the restrictions the Russians imposed on themselves . . . and they cost the Russian space scientists dearly.

For instance: One of the great early scientific discoveries of the American space program was the existence of a hitherto unsuspected layer of charged particles that surrounds our planet. That belt was first identified and described by an American, James Van Allen, and so it is named after him as the Van Allen Belt.

But the American spacecraft that Van Allen was monitoring was not the first to detect the existence of the belt. That feat was, in fact, accomplished by the very first Soviet satellite to go into orbit, the tiny *Sputnik*. *Sputnik* had its own radiation detectors and its own radio to report what it was detecting, supervised by Van Allen's Soviet opposite number, a Russian scientist named Orlov.

The reason it isn't called the Orlov Belt is that Orlov was playing in hard luck. He never found out what his *Sputnik* had observed.

Sputnik faithfully radioed back its observations to Earth, but Soviet paranoia had dictated that its reports be transmitted in a secret military code. That meant that for most of *Sputnik*'s orbit its messages were wasted, since no non-Soviet could read the code. It was only when it was over Soviet territory that its readings could be received by anyone who could understand them—but then it was at the low point of its fairly eccentric orbit. It was impossible then for *Sputnik* to report the existence of the belt . . . because at that point it was no longer in it.

AT THIS MOMENT of writing, however, the worst problems of the Russian space program are no longer either geography or paranoia, they are social and economic chaos. The space scientists and engineers who were once among the elite of the Soviet state are now in parlous state. They have seen inflation slash their salaries almost to invisibility, and they can't always collect even that pittance. Scientists and engineers may stay on the job anyway, out of professional pride, but ordinary workers are not so motivated. Consequently nothing is being maintained or repaired, and much of Tyuratam and the other spaceport at Baikanur is falling apart.

That is a tragedy for the world. Still, it is not to say that there is nothing for the science tourist to see in Russia. For instance, there's Star City.

STAR CITY IS only an hour's drive from Moscow, but for most of its history very few foreigners were permitted to make that drive. When I was there, in 1987, Gorbachev's perestroika and glasnost were just beginning to make themselves felt. Nevertheless, it still took special permission for an American to be allowed to visit.

Star City is where the cosmonauts train; part of their training is parachuting, not so much because they will ever have any call to do it in the conduct of their space missions but as part of the general toughening-up regime. Indeed, on the way out to Star City we saw chutes blossoming under a Soviet air force transport off over the hills.

There were giant radars on those hills, too—to track the spacecraft, my Russian friend said, but I have a notion they were something else, as well. By years-ago treaty the United States and the Soviet Union allowed each other to construct one antimissile system apiece. The American one was abandoned after a while as useless, which it was; but the Soviets kept theirs in operation to protect the city of Moscow, and those radars looked to me like a part of it.

Star City is in fact a whole little city, created from scratch for the Soviet space program. The cosmonauts and their families live there during training, in comfortable small houses; but its heart for the tourist is the space museum. It contains much the same things as the American ones in Houston and Huntsville and the Cape, the only significant difference being that these are Russian. Among other things the Star City museum contains many of the actual space suits worn by the first generations of cosmonauts. It used to have the space suit belonging to Alexei Leonov, a cosmonaut worth remembering if only because of his adventurous return to Earth from one early mission.

When Leonov and his partner, Pavel Bolyayev, reentered the Earth's atmosphere, something went wrong with their automatic controls. They had to guide themselves down by emergency manual control, and when they touched down they found themselves a thousand miles from their expected landing position, in the middle of the great, wild Siberian forests.

They didn't think they were in great danger. Soviet mission control knew where they were. Rescue teams had been dispatched at once. But while Leonov and Belyayev were waiting for the rescue crews to arrive, they built a little fire to keep warm.

That was a bad idea. It attracted the attention of a huge Russian bear, which came snortling and charging into the clearing where the cosmonauts sat. The intrepid cosmonauts leaped back into their capsule and slammed the door . . . and then they sat there, cold and cramped, for another twelve hours until the rescuers arrived.

You note I said the Star City museum *used to* have Leonov's space suit. It doesn't have it any longer. The cash-starved Russian space agency turned Leonov's suit, along with a lot of other trophies, over to Sotheby's in New York, where they were auctioned off to raise money. So if you'd like to see it now, you must first make friends with its new owner, who picked it up for a mere $255,000.

The Star City museum had other interesting space suits when I was there, such as the one that belonged to the first Soviet woman in space, Valentina Tereshkova; it looked exactly like all the others, except that it was a trifle smaller, and someone had embroidered little flowers on it. I had other questions, but about that time a retired cosmonaut came out to greet us, very imposing in his Soviet general's uniform, and it didn't seem polite to pursue the question of that space suit's personal plumbing.

Now Star City is open to almost everyone who has the price, and what remains of the Intourist agency will be glad to sell you a day trip from the city of Moscow. But you don't have to leave Moscow to see some fine mementoes of the Russian conquest of space.

The best place to stay in Moscow for that purpose is in the Hotel Cosmos, huge and—for Moscow—astonishingly tourist-friendly. (It even has a bowling alley and a swimming pool!) What's more, it is located right across the street from Moscow's own vast Space Museum, which has everything you would expect to find in it, including an actual lunakhod, one of the two giant, wire-wheeled buggies the Russians designed to travel across the surface of the Moon. One lunakhod actually got there and is still sitting dead and useless on the Moon's surface because contact was lost; the other is in the Moscow museum.

Even the street you cross to get to the museum, the Avenue of the Cosmonauts, is worth a visit. There's a separate monument for each of the great ones along its length, usually (or it used to be usually) supplied with fresh flowers at the base of the monuments dedicated to the astronauts who had died. At the head of the avenue is a tall, chromium-steel curved metal shaft. It is intended to represent the exhaust of a spacecraft at launch, and around its stone foundation is a frieze depicting every Russian space pioneer, from Konstantin Tsiolkovsky, to Yuri Gagarin, with (unless some Russian president has since had it chiseled off) the ghostly countenance of Vladimir Ilyich Lenin floating benignly above.

PROBABLY THIS WOULD be a good place for me to mention that, if there is any group of human beings in the world I mordantly envy, it is the world's corps of astronauts and cosmonauts. Nor does it make the envy less bitter to discover that they are among the world's most likable and decent human beings.

Between the astronauts and the cosmonauts there isn't that much to choose, either; apart from language, the major detectable difference between a cosmonaut and an astronaut is that the cosmonauts are on average a few inches shorter. Perhaps that is due to diet or genetics;

my personal theory, however, is that it is because their earliest spacecraft were tinier even than the Mercury and Apollo capsules and the tallest cosmonaut cadets had to be selected out.

One of the great pleasures of the years I've spent chasing science around the world is that I've had the privilege of spending time with some of the these marvelous people—with Edgar Mitchell, for instance, the astronaut who managed to squeeze out enough personal time while on the surface of the Moon to attempt to send telepathic messages to a friend on Earth. (The experiment didn't work—perhaps because the exigencies of his work schedule made him miss the appointed time for the experiment—or perhaps not.) And with Michael Collins, the man whose fate was to circle in orbit while the very first of all lunar-landing astronauts, Neil Armstrong and Edwin Aldrin, were making history on the surface of the Moon far below.

Collins didn't have the most enviable job in NASA history. For the twenty-one hours and thirty-seven minutes that he orbited the Moon alone, while Armstrong and Aldrin were gallivanting around on the lunar surface, Collins's job was merely to wait for their return. As he orbited he wore around his neck a pouch containing instructions on emergency procedures to be undertaken at once if certain things went wrong. For one possible problem, though, there was no prescribed procedure. In the event that the lunar landing module simply was unable to take off again, Collins had no orders to do anything in particular. That was because there was nothing he could do. It would have been impossible to take the orbiter down to the surface to rescue them; and during all that weary time of waiting, Collins was contemplating what it would be like if the takeoff failed and he had to abandon his comrades there to die, as he headed back home alone.

IN THE OLD Soviet days I had no chance to visit Star City or to have many one-on-one conversations with a Soviet cosmonaut. Those superstars were on view from time to time at special occasions, like American sports figures or movie stars, but that was as close as I ever got. Then things loosened up and in the 1960s I got my chance to spend some time with a cosmonaut. His name was Vitaly Sevastianov.

Sevastianov never reached the Moon, but he almost did; he was one of the six cosmonauts named to the short list for the first Soviet round-the-Moon flight. What kept Sevastianov from getting there was the failure of the Soviet equipment. The lunar flight was postponed time after time because of failures of their Proton heavy-lift rocket or of their lunar capsule itself; and then, after *Apollo 8* orbited the Moon on Christmas eve of 1968 and the Soviets could no longer be first, the mission was scrubbed entirely.

Still, Sevastianov achieved a remarkable personal list of achievements; at one time he owned the record for accumulating more hours in space than anyone else in the world. He was, like all the other cosmonauts and astronauts, a pampered favorite son of his country. He had been given everything his country had to offer, including a spacious apartment on Gorky Street, the Fifth Avenue of Moscow, with a solid wall of trophies and medals, and even a little red Italian sports car. It was an experience to tootle around Moscow in that car with Vitaly Sevastianov. Even car-short Moscow had its traffic jams, but not for a hero cosmonaut. Sevastianov had the privilege of driving in the VIP central strip, called the Kremlin lane, that was on every main Soviet road. Besides, every traffic militiaman in Moscow recognized his car and stopped traffic to let him through. And while we were having drinks in his apartment (French brandy, of course; no Armenian stuff for a cosmonaut), his young daughter's puppy came in to investigate us. Sevastianov wrote out the dog's name for me: СиУпIч. Which, if you can make out the Cyrillic alphabet, you will recognize as the phonetic equivalent of Snoopy.

The Soviet cosmonauts were spoiled, all right—just as our own early astronauts were, with their free houses and financial perks—but who can object? To be sure, the astronauts and cosmonauts were not the true heroes of space exploration; that honor belongs collectively to the hardworking geniuses who designed and built the spacecraft and the tens of thousands of workers who supported them. But the astronauts and the cosmonauts risked their lives—and some lost them—and most of all, they were the ones who chased science as far as it is now possible to go. They were Out There, and I would change places with any one of them in a hot minute.

PUTTING HUMAN BEINGS into space is the most theatrical part of the world's space programs, but not the most important. The real work of space exploration is done by machines, unmanned spacecraft that can go to places far out of the reach of present manned ships.

What unmanned spacecraft do is observe. Almost all the kinds of astronomical observatories we have on the surface of Earth are duplicated in spacecraft—infrared and ultraviolet, radio, X ray, gamma ray, visible light, whatever. A spaceborne telescope is seldom as big as any college's campus instrument, but it doesn't have to be. The ones that bring us the best pictures ever of our own Sun's family of planets get right up close to the planet or moon they want to photograph.

Then they send their pictures back to Earth; and the place where most of those pictures come to is in Pasadena, California, and it is called the Jet Propulsion Laboratories.

AS YOU CAN tell from the name, JPL originally did have something to do with jet propulsion, back in the World War II days when it was founded. Those days are long gone. Now it is run by the California Institute of Technology, and it has become the world's premier center for receiving, analyzing, and interpreting data from space probes.

Doing this involves specialized duties of many kinds, one of the most important of which is running what is called the Deep Space Network. The radioed reports from American spacecraft need a receiver on Earth, and a big, sensitive one at that. Since the Earth revolves and thus does not always present the same face to the part of the sky inhabited by the spacecraft, more than one receiver is needed.

That's where the Deep Space Network comes in. Three main DSN receiving stations are spaced around the world, so that one of them is in position to get incoming data at all times. These stations are in Goldstone, in California's Mojave Desert; in Madrid, Spain; and in the little Australian town of Tidbinbilla, not far from Canberra. What each looks like is a backyard satellite antenna, but vastly enlarged. Each main dish is more than two hundred feet across. It can be turned in any direction, like a great optical telescope, and, mount and all, each unit weighs around three thousand tons. Since the signals that come from the farthest and weakest of our spacecraft are almost vanishingly faint, there's a huge and sophisticated amplification system built into the package. Once received, the signals are transmitted instantly to JPL in Pasadena.

JPL's physical plant is located on a rolling campus amid green hills just outside the town of Pasadena itself. Of all the science I've ever chased, some of my very best trophies were caught right there at JPL.

For my first sight of JPL's planetary snapshots I wasn't actually in Pasadena. I was in Boston, taking part in a conference with Isaac Asimov. During breaks we hurried up to my hotel room to catch the latest photographs on the TV set. The occasion was the first close flyby of Mars, and as we saw those precious first photographs of the Martian surface we looked at each other in shock. "Craters?" Isaac moaned. "How come neither of us expected *craters* on Mars?"

I didn't have an answer for that, but I did make a resolution. I resolved that one way or another, next time there was a good planetary flyby I was going to be in JPL myself to see everything that came in at the moment it arrived.

WHAT DO YOU get at JPL that you can't get on your TV set or computer? Well, you get more of it: the best of TV specials shows only a selection of the pictures that come in from the Pioneers, the Voyagers, and the rest of our space probes. You also get it *first*. You can be there when the transmissions are relayed direct from Goldstone and, for the first time ever, reconstituted as television images. You are seeing what no human eye has ever seen before you.

So next time there was a major flyby, I made sure I was at JPL to watch it.

I didn't really expect a whole lot of new data. The occasion was a *Mariner* flyby of the planet Venus, and, really, Venus isn't much to look at no matter how close you get to it. In the optical frequencies all you can ever hope to see is the dense cloud cover that permanently blankets the whole planet.

But that, of course, only whetted my curiosity. In 1967 Venus was just about the greatest mystery in the solar system. It would seem that it shouldn't have been a mystery at all, because it was the planet that came closest to Earth, but there were those clouds. No one could say what Venus's surface was like. There was a notion for a while that Venus was simply a younger, hotter Earth, possibly jungly with hot swamps, maybe even with bellowing dinosaurs. That was a fun model, and it gave story material to dozens of science fiction writers from Otis Adelbert Kline to Stanley G. Weinbaum.

But, though more careful analysis had shown that that model was definitely wrong, there was no clear consensus about what was right. Some astronomers believed that under the clouds was nothing but hot, dry, bare rock. Some, on the contrary, thought the whole surface might be covered with a vast ocean of water. A few speculated that there might be an ocean, all right, but that instead of water it would be composed of oil.

So there were a great many questions in my mind as I flew to California, and I hoped that *Mariner* would, well, not answer them;

that was much too much to hope for, but at least supply a few new hints.

It didn't work out exactly as I'd hoped. Oh, I did get some interesting new data, but it didn't come from *Mariner*.

It wasn't that anything went wrong with the mission. *Mariner* didn't fail. The spacecraft flawlessly did what it was supposed to do: passed by Venus within three thousand miles, just as close as it was intended to, and made all the observations it had been programmed for. But the joy of the dozens of scientists gathered to observe their triumph was restrained. The very day before the *Mariner* flyby the confounded Russians had *landed* one of their Venera spacecraft on the planet, and *Mariner*'s comparatively modest success was as ashes in the mouths of the Americans.

By the way, the Venusian surface turned out to be hot and dry.

Well, things got better after that. Over the next twenty years American spacecraft explored Venus many more times, as well as Mars, Jupiter, Saturn, Uranus, and Neptune. For many of those great flybys I was right there in JPL's Von Karman auditorium, hanging out with the rest of the space freaks as we watched those glorious pictures arrive.

In the earliest days there was a special thrill to see them coming in. The data transmission rate was slow, and the computers of that generation couldn't handle the task of assembling them in color before they were displayed. So the pictures appeared on the monitors in black and white, one line at a time building up on the screen before our eyes—pictures no other human eye had ever seen! For the Jupiter flyby they were truly astonishing: the Great Red Spot as it had never been photographed before, huge and swirling. The rings of Jupiter, which no one had ever before even suspected. And, most of all, the astonishing variety of Jupiter's family of moons, each one as different as the Sun's planets themselves.

For Saturn there was the unanticipated structure of the rings. In your backyard telescope the rings look like nothing more than a bright plumbing gasket, if you are lucky enough to see them at all. Even Palomar's Big Eye shows only a few of them, separated by empty spaces like the one that is called the Cassini division. But the spacecraft see more. There are at least a hundred thousand discernible separate rings, some of them strangely scalloped or braided, all of them weird. What keeps them in order is the gravitational attraction of Saturn's moons, including some tiny ones, never seen before our spacecraft arrived, that are embedded within the ring system itself.

When the first of those new moons turned up on the JPL monitors, a couple of the resident astronomers wondered who would get the privilege of naming it, which led immediately to proposals for what it should be named. My friend and fellow spacelover Jerry Pournelle piped up, "Considering that it is shepherding the rings, there's only one right name for it. We ought to call it Sheepdog." So they should have; but more sedate opinions prevailed, and it is now known as Prometheus.

THEN I MADE a big mistake. I skipped the Uranus flyby. I figured there wasn't likely to be much to see in that rather dull planet, and I had other things that seemed more important to do at that time.

I was wrong. I missed being on the scene for all the startling new discoveries about the Uranian moon system, particularly that most bizarre moon of the lot, Miranda. No human being was prepared for the way that one looks. It is a patchwork affair, looking as though some cosmic Dr. Frankenstein had put it together from salvaged spare parts. (Probably that's exactly what the laws of physics did. Miranda was most likely a moon that got too close to giant Uranus, was torn apart by the planet's gravitation, re-formed into a single body again, was torn apart—perhaps half a dozen times in the history of the system, and quite likely facing the same prospect at some future time again.)

That taught me a lesson, and so I made it a point to be there for Neptune.

Of course, the star of that show wasn't the planet Neptune itself, it was Neptune's wonderfully weird major satellite, Triton. Every picture that came in was a new wonder or a new puzzle: oceans frozen solid, wind streaks left on the surface—by an atmosphere thinner than in the lightbulb over your head; one pole bright with—one supposes—water ice and the other nearly black with—one can only guess—some other kind of ice that no one has ever seen before. Not to mention the giant geysers that produce fountaining plumes of—it seems—some sort of liquid, on a satellite in the coldest part of the solar system, where liquids of any kind are scarce.

The flyby events at JPL have changed in character since the early days. Now the computers process the data almost instantly and the pictures come up on the monitors all at once, in color. Now a lot of other people have caught on to the excitement of being there, so the crowds are huge; they sell T-shirts, posters, and photographic prints in stalls on the way to the auditorium, and the whole event is like a county fair. But it's still wonderful.

Even if that particular part of it is pretty much over.

There's still one planet left to explore—tiny, remote Pluto—but now funding is scarce and priorities are elsewhere. So I do not count on getting that precious first close-up look at a new planet any time soon.

But there are comets and asteroids . . . and even if we finally do manage to use up all the explorable objects in our solar system, there are farther horizons. It's a *big* universe.

Actually, we've already begun to explore at least the fringes of it, because some of our earlier space probes—built and launched in that pre-shuttle, pre–Star Wars defense time when NASA could afford to construct spacecraft that would last—are already doing so.

If you can locate the star Aldebaran in the sky, look in that direction. What you're looking at in addition to the star—don't expect actually to see it; it's too tiny and too far away—is the first interstellar spaceship of the human race, *Pioneer 10*.

Pioneer 10 was launched on March 2, 1972, to give us an early look at, among other things, Jupiter and its moons. It didn't stop at Jupiter. It didn't stop at the edge of the solar system at all. It kept right on going.

When last heard from, in 1997, it was out in the remote Kuiper belt of comets, more than 6 billion miles from Earth, and far beyond any of our planets. *Pioneer 10* won't be heard from again. It's powerless and silent now, but it's still heading toward the neighborhood of Aldebaran, sixty-eight light-years away. At its current velocity of six miles a second, half a million miles a day, it should get there in—oh, say about 2 million years.

If there are any spacefaring aliens out there, perhaps one of their ships will find it and take it aboard; and if they do they'll find a message from us to them.

The idea of putting an actual letter in the mail for E.T. originally came from an English science journalist named Eric Burgess and a young jack-of-all-technology named Richard Hoagland. Burgess and Hoagland took their notion to that father figure in extraterrestrial research, Carl Sagan, and Carl recognized a great idea when he saw one. He took it from there. He persuaded the artist he was married to at the time, Linda, to sketch a message, and he convinced the people at JPL that it would add nothing to the weight of *Pioneer 10* to engrave it on the spacecraft's side. So they did, and it's on its way.

It portrays a pair of human figures to show what we look like, including the fact that we come in two slightly dimorphic genders. Lines radiating away from them figures show where we are: Each terminates in a pulsar, and the length of the lines shows how far we are from each of them. So E.T., if he is clever, will be able to figure out not only who sent the message but where we're located in space.

There are some people, some scientists among them, who really didn't want that message on the side of *Pioneer 10*. They thought the human race might some day come to regret it. What if aliens do some day find that letter, along with its directions for finding us—and they are the kind of murderous, ravening, ray gun–wielding monsters the early science fiction pulps contained in every issue? Do we really want these people to find us? Would we be sorry we'd mailed them this road map?

Maybe so . . . but, considering how long this form of spatial delivery takes, it is not, I think, anything that we have to worry about just now.

Chapter 5
WHEN THE EARTH MOVES

Volcanoes and Earthquakes

Sometimes you don't have to go looking for examples of science in action. Sometimes they thrust themselves right in your face.

For instance, there is the incident that occurred in 1989. On the afternoon of October 17, like many another American, I flipped on the TV to catch the World Series game between the Oakland Athletics and the San Francisco Giants. What I got was a considerable surprise. The screen didn't show the familiar confines of Candlestick Park. Instead, it showed a slide bearing a brief and uninformative message:

WE HAVE TEMPORARILY LOST OUR SIGNAL
Please stay tuned

The signal kept right on being lost, too, With all the redundancy in communications available to a major television network, that sort of thing simply does not happen. So I stared at the screen in perplexity for a few minutes, then called to my wife in her own office down the hall, "Something big is happening in San Francisco. Do you suppose they could be having an earthquake?" And, of course, they were.

It didn't take long for the networks to catch their breath and deploy some of those vast communications resources. Connections were made, helicopters were sent up, and by bedtime I had seen all I wanted to see of a section of the great Bay Bridge collapsed upon itself, the lights of the rescue workers along the formerly two-level, now pancaked, stretch of Highway 880 in Oakland, and the blazing fires all around the Marina section of San Francisco itself.

That one—it is called the Loma Prieta earthquake, because that's where its epicenter was (and I'll tell you what an epicenter is in a moment)—was the first really bad earthquake any city in the United States had experienced in my lifetime. The last major one was the quake and fire that had obliterated most of downtown San Francisco in 1906, eighty-three years earlier. It seemed fairly reasonable to think that it might well be another eighty-three years before there was another. It wasn't, though. It was only five before it was Los Angeles's turn. That was before dawn on the morning of January 17, 1994, and by the time the Sun came up highway overpasses had crashed to the ground, broken gas pipes had started unquenchable fires, and some sixty-one people died.

WHY ARE ALL these big ones in a single state?

Simple. California is Earthquake Heaven. Geography makes that inevitable, for the state is located at the exact point of crunch where two huge tectonic plates are grinding against each other. That's a hair shirt for Californians to wear, but it does make for some interesting spectacles for us science watchers to look at.

It's more interesting to watch when we know what it is we're looking at, so, if you don't mind, let's talk a little bit about the science underlying earthquakes, which is called plate tectonics. (If you do mind, either because you already know all that or because you don't care, no problem. Just skip to the next break in the text.)

The science of plate tectonics is pretty new as scientific disciplines go, although there were suspicions about it as far back as the year 1620. That was when that bright and curious-minded man, Francis Bacon, studied the new maps of the landmasses around the Atlantic Ocean that had just begun to be made by far-voyaging European explorers. Bacon noticed a funny coincidence. Queerly, the eastern coast of South America looked as though it would fit right into the western coast of Africa, if only the broad Atlantic Ocean hadn't happened to lie between them.

That was certainly an odd fact. No one thought it was a particularly important one, though, for no one was rash enough to suppose that continents could actually move apart . . . at least, not for the next three hundred years.

Then, early in the twentieth century, a man named Alfred Wegener came along and said that the fit between the coastlines was no coincidence. He said it was because, yes, the two continents *had* once fitted together, and the only reason they were apart now was that all the solid land of the planet was in constant slithering motion around the world. He called that process continental drift. Not many scientists paid any attention, though, for reasons that had as much to do with the somewhat unlovable personality of Wegener himself as to the merits of his ideas, and Wegener's speculation was pretty quickly forgotten.

However, it was rescued, a generation or so ago, by a Canadian scientist named J. Tuzo Wilson. It wasn't just continents that drifted, Wilson proclaimed. Every last bit of Earthly crust—continents, islands, sea bottoms, and all—was floating on the planet's semiliquid core of magma, like patches of fat on the surface of a pot of cooling chicken soup. These rafts of crust, called tectonic plates, didn't move rapidly—your fingernails grow about as fast as they travel—but they did continuously keep on moving in more or less random and sometimes conflicting directions.

Sometimes these floating plates were pulling away from each other. In that case new low-lying crust was formed to fill the gap between them. Then some of the water from the world's seas flowed into that depression to make an ocean. Such an area of that sea-bottom crust had been filled in that way, long ago, with what is now the Atlantic.

Sometimes the plates collided head-on. Then mountains were pushed up where they struck, as one plate was forced down underneath another—was "subducted," as they say—and thus lifted the other one higher with what a teenager might call a kind of superwedgie. That is how the small plate containing India slid northward to strike the huge plate that held Asia and throw up the Himalaya Mountains.

Sometimes two plates merely ground against each other as they moved in opposite directions along their common border: thus, for instance, what is now going on in California.

The more scientists study the evidence, the more it shows that Wilson had the right of it. The Atlantic Ocean is in fact growing and has been growing for millions of years. The split between South America and Africa began 80 million years ago and is still widening. The evidence is there. Drill ships have studied and dated cores from all over the Atlantic bottom, and the farther the cores are taken from the oceanic center line that is the source of the spreading—it is called the mid-Atlantic Ridge, and I'll have more to say about it in a moment—the older they are. Along the ridge itself new flows of lava are continually welling up to push the old crust aside. If the continents of Africa and South America look as if they had once touched each other, it is because they indeed did. The proof is that geologists and paleontologists have shown that they share the same mineral regimes and the same fossils right up to the time of separation, after which the species begin to diverge.

Seismologists have identified more than a dozen vast tectonic plates. There is one for almost the whole Pacific Basin, another that holds North America and part of the Atlantic; a third, the Eurasian, meets the North American in midocean along that aforesaid mid-Atlantic Ridge; others for Africa, for Australia (along with India and much of the seas between), for South America along with its share of the South Atlantic—for every part of the Earth's surface. All of these plates are in motion.

And California lies just where the great North American plate, moving more or less in a southward direction, rubs against the even greater Pacific plate, moving more or less north.

The masses do not slide easily against each other. They stick for a time, building up strain, and then they slip; that sudden release of strain is the earthquake.

California is riddled with "faults" where the strains accumulate, the most famous of the faults being the notorious San Andreas. The San Andreas fault is one of the longest of them, too, threading through the heart of the state from near the Mexican border until it reaches San Francisco Bay. There it strikes offshore into the Pacific.

Although the real fault activity takes place underground, faults often show themselves on the surface. The San Andreas fault isn't only visible, it is sometimes quite conspicuous. If you fly over it in a light plane you can see places where westward-tending streams jog to the north as they flow toward the sea, their courses offset by the movement of the plate. Or, if you want a look at the surface mark of the fault itself, a good place to go is the city of Palm Springs. There you can take the cable car to the top of the mountain just outside the town. From the mountaintop you can see a clearly marked line drawn across the scenery to the west. That's the San Andreas.

Actually, most of California's earthquakes arise from some fault other than the San Andreas, not because it isn't the biggest but because there are so many more of the smaller ones.

The slippage that spoiled the World Series game in 1989 took place in a small and inconspicuous fault many miles south of Candlestick Park and a couple of miles underground. The point on the Earth's surface directly above where such a slippage occurs is called the epicenter, and in this case it was in Loma Prieta, California.

* * *

IT WASN'T QUITE. Most of the city's buildings stayed erect, even the skyscrapers; they had been built to survive a quake of that size. The buildings that didn't survive were almost all the URMs—the unreinforced masonry buildings—that stood in the city's Marina district. That whole area was made land. There what had once been shallow coves of the bay had been filled with debris to provide new real estate to build on. It is not good to be in a URM on made land during an earthquake. The vibration causes the fill to liquefy, and supports do not hold.

When I was last in San Francisco, four years after the event, there was not much left to see. The Bay Bridge was repaired. Highway 880 was working again, but no longer double-decked—it had been a nervously supple, bouncy road to drive on anyway. Most buildings were repaired, though a few, even including San Francisco's City Hall, were still propped with timbers. And the city was waiting for the next one.

ALTHOUGH EARTHQUAKES ARE most common where tectonic plates crunch against each other, they can happen anywhere.

The state of New Jersey, for instance, is more than a thousand miles from any plate border. Still, one afternoon when I was standing in my kitchen near the town of Red Bank, I felt our house shake as if a very heavy truck were rumbling past. There wasn't any truck, though. It was an authentic earthquake. By California standards it was a pitifully small one, to be sure, no more than a gentle Force 4 or so . . . but in that area, where I had never felt an earthquake before, it was pretty startling.

The very worst earthquakes ever recorded in American history took place just as far from any plate border. They happened in the years 1811 and 1812 and there were three of them in a row—gigantic convulsions, estimated at 8.6, 8.4, and 8.7 on the Richter scale. Their epicenter was around the tiny settlement at New Madrid, Missouri, and their force was almost unimaginable. The movement of the ground was almost a hundred times as violent as the 1994 Los Angeles quake. In New Madrid the shocks uprooted whole forests and shook buildings flat—what few buildings were in the area. They made the channels of the Mississippi River itself run backward for a time and changed its course permanently in some stretches. "The whole land was moved and waved," one eyewitness wrote, "like waves of the sea." Another said that after each great shock, all night long, he was kept awake by the noise of brand-new cliffs, heaved many feet high by the quake, crumbling and falling into the brand-new streams.

Going on two centuries later, you still can see some of the awesome effects of the New Madrid event from the air in the shape of offset stream courses. Few injuries or deaths resulted from it. But that was only because there were relatively few people living in the area then, and almost no large buildings for them to be trapped in or crushed by.

Will the New Madrid fault strike again? Seismologists think it likely; their digging has indicated that such an earthquake strikes that area every few centuries. And if one came now, the toll would be far worse than anything Americans have ever experienced. Now more than a dozen great cities, from Memphis to St. Louis, would be well within the impact zone, and its effects can only be imagined.

You will have noticed that I keep on speaking of cities. The reason is that it is the cities themselves that turn earthquakes into mass murderers. Nobody dies of an earthquake all by itself. At the worst the ground shakes, you fall down, maybe you have a moment's seasickness, and that's the end of it. That is, that's the way it is provided you are out in the open when the quake hits. But few Americans spend much of their lives completely away from buildings, dams, or bridges. What people die of now is being caught in the tumbling walls of the homes or offices they inhabit, or from the floods or fires that follow, or of the diseases that come when sewage and water systems are destroyed.

I've talked about American earthquakes, but there are plenty of earthquake-prone regions elsewhere on the planet. Worldwide, there are about a hundred detectable quakes every day, with a big one—say, a 7 or so on the Richter scale—coming along about once a month. While this book was being written there were big ones in Armenia, Italy, China, and Turkey, among other places. There was a not-so-big (but big enough) one in Japan, which is maybe even more earthquake-prone than California. (The reason for that is simple enough. The Japanese islands lie over the junction of three competing tectonic plates—the vast Eurasian and Pacific plates and the smaller Philippine one. Quakes are a natural result.)

Where will the next really catastrophic earthquake happen?

Nobody knows that. Still, if there were an office pool going on the subject I know what I would pick. It would be Beirut, Lebanon.

That's not because Beirut has been hit much recently. It hasn't.

There *may*—records are sketchy—have been a terrible quake in the year A.D. 551; that is, something really catastrophic happened then, demolishing the city, but historical records are not clear on what the catastrophe was. But Lebanon sits right on top of what is called the Dead Sea Transform, where the Arabian tectonic plate meets the African. Over the last 18 million years the two plates have slid nearly sixty miles past each other, moving at the rate of about a quarter of an inch each year. They're still sliding, and in my personal opinion (backed by no authorities at all) that bodes ill for the mammoth construction of high-rises now going on all over Beirut, after the devastation of the recent wars.

There is even speculation that around 1200 B.C. something similar happened to the cities of the Fertile Crescent, where writing and laws and many other attributes of Western civilization began. When archeologists excavate such cities as Troy and Mycenae, they find many bodies crushed inside buildings, and what beside an earthquake kills in that way? All those nations disappeared from history at about the same time; and it may be that repeated quakes were what wiped them out. In the Western Hemisphere something of the same sort may account for the collapse of the civilization of the Mayas.

Sometimes, rarely, earthquakes have a comic side. Some years ago when I visited the city of Skopje, capital of the Macedonian region of what was then still Yugoslavia, I found that the people had turned their recent earthquake into a tourist attraction. Over the town hall square, where Skopje's young men and women paraded every Saturday night in concentric, counterrotating circles, the town clock was stopped at the time of the quake.

It was an interesting souvenir, but unfortunately a fake. The clock hadn't been stopped by the quake. It had been broken and left unrepaired for some years. It was only after the quake that some enterprising Skopjean had climbed the building and made it into something for tourists to photograph, by resetting it to the time of the event.

WE'VE GOTTEN A little bit away from things you can go and see, but such things do exist. For instance, there are seismographs.

A seismograph measures the shaking of the earth. In a museum in Beijing I saw an early version of one that was nothing more than a ring of outward-facing jade frogs. Their wide froggy mouths were open, and each frog mouth held a little porcelain ball. When the ground shook some of the balls would fall out. The strength of the quake could be gauged by how far they fell and its general direction estimated by which of the balls fell.

Most working seismographs are kept out of sight in working geological laboratories, but some are open to the public. If you're picnicking in Los Angeles's Griffith Park, you can go right up the hill to the Griffith Observatory and see some in operation. These are the classic, but now old-fashioned, scientific type: rotating drums of paper, with a suspended needle scribing a line. When the ground shakes the drum moves. Inertia causes the spring-suspended needle to move more slowly, and so it traces jagged peaks and valleys. The bigger the peaks and valleys are, the stronger the quake, and by comparing arrival times at various seismographs located around the area, the source of the earthquake can be located. (More recently seismographs have become electronic, like everything else, but the principle remains the same.)

MAYBE I'VE SAID enough about earthquakes, because there's a lot more science to cover. For instance, that underground sea of oozy magma that we float atop on our tectonic plates does not always *stay* underground. From time to time, in one place or another, some of that molten rock breaks through.

Then it becomes what we call a volcano.

I've had a personal weakness for volcanoes ever since the tail end of World War II, when I happened to live on one for some months. It wasn't just any old volcano, either. It was Mount Vesuvius, perhaps the most famous volcano in the world.

The reason I was there was that halfway up the mountain there was—still is—a hotel named the Albergo Eremo (translation: Hotel Hermit). By deft footwork the advance elements of my unit, the 12th Weather Squadron, USAAF, had managed to beat out all other claimants in the race to requisition it as a rest camp for us highly stressed weathermen.

As a site for R&R, the Eremo was first-rate. The postwar explosion of cars and people had not yet turned the region's air into smog and the Bay of Naples into sewage. From the Eremo's wide verandas you could gaze out upon one of the most spectacularly beautiful vistas in the world, Naples off to the right, Sorrento and the Isle of Capri to the left, and the stunning bay itself lying in between.

If the Eremo had a flaw, it was that getting up to it involved miles of hairpin curves and guardrail-less climbs on road surfaces of shifting crushed pumice. (Half a century later, the road has been paved—more or less—but it is still a fun drive for the adventurous, a disconcerting one for the rest of us.)

To be sure, the old mountain did twitch now and then, and just a few months before I got there it had done more than twitch. The lava flows from that 1944 eruption destroyed two villages and buried many acres of farms and vineyards as well as a good many farmhouses and the funicular railway that had been the sensible alternative to driving that mountain road. There was loss of life, too. The eruption killed twenty-six people, and while it was doing all that it seriously worried the American and other Allied troops who were chasing the German army up from Salerno. (American servicemen's anxious relatives, listening to radio reports of how things were going in Italy, got news of actions on three fronts: Anzio, Monte Cassino—and Vesuvius.)

The Hotel Eremo was in little danger from that eruption, though. The people who built it wisely situated it near Vesuvius's vulcanological observatory. Those nineteenth-century seismologists knew their mountain, and so they had been careful to locate their installation on a hump that rose from the volcano's flank. Lava streams might roll down the mountain toward them, but they would almost certainly divide and run around the mound that contained the observatory and the hotel. Getting out of the hotel safely during an eruption might well be a problem; getting fried probably would not.

You can drive a little farther up the volcano than the Eremo, but if you want to look down into the crater the last mile or more has to be on foot. It is not an easy climb. I was willing to undertake it when I was twenty-four, but I haven't been willing to do so more recently. (There are charter helicopter flights available for those who want the view but not the exercise.) Generally there has not been a lot to see in recent years, because Vesuvius's crater appears to have been cooling down lately. Apart from the occasional small shudder and the puff of cloud that often hangs over the peak, Vesuvius might seem extinct.

That appearance is not to be trusted, though. A half century of quiescence means nothing in the million-year life span of a volcano. In the time of the Romans they thought the mountain dead (just as the people of the West Coast did of Mount St. Helens before 1980), but a little after noon on the twenty-fourth of August in the year A.D. 79, the Romans learned better. That was when Vesuvius blew its top.

Fortunately for all of us later spectators, a trained reporter was on the scene, that jack-of-all-arts known to history as Pliny the Younger.

Pliny, who saw most of the eruption from a boat offshore, wrote a full description of the event in two long letters to the historian Tacitus, and they have survived to the present. At first flames erupted from the volcano. Then a ten-mile-high black cloud appeared, laced with lightnings and shaped, Pliny said, more like a pine tree than anything else he could think of. Steadily, remorselessly, the cloud marched on the pleasant little provincial city of Pompeii, and it dropped its burden of poisonous gases and incandescent ash right onto the town.

Twenty thousand people lived and went about their business in Pompeii that morning. By afternoon nearly every one of them was dead, poisoned by the gas or broiled and crushed by the fiery fallout. The ashfall covered everything, smashing roofs down, toppling pillars and statues, to a depth of a dozen feet and more. Just down the coast the little Roman port city of Herculaneum died that same day, though in a different way: It was mudslides from the slope of the mountain that buried Herculaneum, but the death toll was the same. If you were there, you died. And both towns remained buried for nearly two thousand years.

Will Vesuvius erupt again in that same catastrophic fashion? There's no reason to doubt it. A couple of thousand years is nothing to a volcano. And—as with the earthquakes in the New Madrid area—with all the increased population and their buildings within sight of Vesuvius, if it came again today it would have an entirely different impact. Augusto Neri, an Italian seismologist from the University of Pisa, reckons that in a similar quake today, more than a million people would die in the first fifteen minutes.

When he was asked what sort of precautions the inhabitants of Naples and Sorrento could take against a recurrence, he replied succinctly: "Move away."

VOLCANOES ARE NOT to be trusted. So the people of Pompeii learned in A.D. 79; so did those who lived near Mount St. Helens, in the American Pacific Northwest, nineteen hundred years later.

I happened to be in Seattle shortly after Mount St. Helens blew on the morning of May 18, 1980, and saw its plume from a distance. (It's actually nearer to Portland, Oregon; forty-five minutes' driving time from Portland takes you to the Mount St. Helens National Volcanic Monument Visitor Center in Castle Rock, Washington.)

As volcanic eruptions go, Mount St. Helens was comparatively small, but still it blasted the top quarter mile off the mountain. It had been 9,677 feet high; what was left was an 8,364-foot stump. The effects of the eruption were vast. Whole forests were smashed flat by the blast. Boiling-hot sludge rolled down the mountain, and virtually every living thing nearby was choked or scalded to death by the flow or by the superhot six-hundred-mile-an-hour winds that went with it. Towns fifty miles away woke to find their streets covered with pumice. Beautiful Spirit Lake, at the foot of the mountain, was filled with ash and downed trees; so much debris filled the lake that its water level was elevated by an astonishing two hundred feet, and all of its fish died. (Now the lake is clean again and back within its shores, but still nothing lives in it. Sportsmen are urging the state to stock it, but scientists are begging that it be left alone as an experimental site to see how natural damages repair themselves.)

The most startling part of the eruption of Mount St. Helens, of course, is that we people who live in the Lower 48 don't expect volcanoes on our home territory. Probably that is why, although the mountain gave signs that it was going to erupt, the local people were reluctant to evacuate. In the long run most of them allowed themselves to be persuaded, but one stubborn holdout, a crusty old hermit named Harry Truman (no, not *that* Harry Truman) refused to leave. His body was never found.

BEFORE WE LEAVE Italy entirely I should mention that there are plenty of other volcanoes around that country. Just down the coast from Vesuvius, on the outskirts of the seaside town of Puzzuoli, is the much smaller volcano called La Solfitara. It has never erupted on anything like a Vesuvian scale, but neither has it ever stopped producing its emissions of boiling water, steam, and smoke. (Its volcanic activity is said to have been the sight that inspired Dante to write his *Inferno*.) La Solfitara also continually produces periodic earthquakes and unsettling rises and falls of the shoreline.

Still farther south lie the volcanoes of Sicily.

As it happens, while I was writing this book I also happened to be staying for a short time in the Sicilian town of Taormina. (It's my habit to do some writing every day, wherever I happen to be.) Every time I looked out of the window of my hotel room, I could see a mountain that filled the southern hemisphere. It was a monster, and it's called Mount Etna.

Etna is the biggest volcano in Europe, its base is 120 miles around, and it is frequently active. When it erupts it emits little ploops of lava from one or another of its many craters. Then the lava slithers downslope until it cools enough to harden and stop—usually without going very far, though from time to time it has buried nearby communities. As a result of this, the final shape of the mountain looks rather like a sand castle after the tide has come in. Etna is far wider than it is tall.

Still, it is tall enough. When last measured its peak reached 11,053 feet, but regard nothing as settled; the volcano's height varies by hundreds of feet as activity raises it and erosion lowers it again. No one could think of Etna as extinct. The American Mount St. Helens has erupted only once in historical times. Even Vesuvius has done so only rather infrequently, but Etna does it over and over again.

It wasn't doing much as I gazed at it out of my hotel window. There was snow on its ridges and a wisp of cloud over its peak—vapors from its crater, maybe, or perhaps just the standing lens of orographic-uplift cloud that hangs over many peaks. (Air moving across the ground is forced upward as it hits the flank of the mountain. Rising, it cools. The moisture it contains condenses in droplets of liquid water—and becomes a cloud; then, as the air drops on the lee side of the mountain, it warms again and once more becomes invisible water vapor. The air and its burden of moisture are always in motion, but the visible part of it, the cloud, seems to hang in place.)

So that day Etna was sound asleep, but when it wakes it is a tiger. It has produced 140 recorded *major* eruptions since the year 479 B.C.; twice it has destroyed the nearby city of Catania in immense lava flows; the number of times it has wiped out smaller communities, on the flanks of the mountain itself, is uncounted. There was a big one in 1983, and it was marked by serious efforts to divert the flow of lava—by bulldozing dikes to try to divert it from its course, by bombing it from the air, even by bringing up fire trucks to try to solidify the lava by playing water on it. The contest was not equal. The lava always won.

Because Etna's slope is comparatively gentle, it is fairly easy to go to the top, first by cable car, then by minibus, and finally on foot; alternatively, there are helicopter flights from Catania that allow you to look right down into the crater.

ALTHOUGH ITALY'S VOLCANOES are the most famous, they aren't the only ones in Europe. Nor was the eruption of Vesuvius that entombed Pompeii the greatest in historic—well, more or less historic—times. For that we need to go to the Greek island of Thera—now called

Santorini—in the Aegean Sea. That's easy enough to do; cruise ships from the Athens port of Piraeus make it a standard port of call, and it is worth seeing.

Santorini began as a single huge volcano that poked itself up from the floor of the sea and became an island. Then it rested. It remained quiescent long enough for the fairly important city of Akrotiri to establish itself on it. Then, one day in the year 1628 B.C., the island exploded.

How big was this Titan among volcanic eruptions? We can't look for nearby eyewitnesses to answer that. None of them lived long enough to set down their observations. Even most of the more distant ones fail us, because their records do not survive. Our best source is Plato, who only heard about the event centuries later, and from Egyptian sources—who themselves had only fragmentary records of what surviving travelers of the time may have seen or may have heard from someone who did see.

Not only is this data sketchy, it is also fairly well corrupted. Either Plato or his Egyptian informants embellished the tale with details, so that in Plato's writings there is quite a pretty picture of the island, its people, and the eruption that destroyed them. Plato called the place Atlantis.

If there ever was a real Atlantis, as distinct from the many legends that have grown up about it, it was almost certainly located on that island of Santorini; and the only surviving "facts" about it come from those few lines of hearsay and conjecture in Plato. Everything else is imaginative embroidery.

What we do have is physical evidence. The record of the rock itself tells us what happened there, and you can see with your own eyes some of its results. When you sail into Santorini's port, you disembark on a narrow stretch of beach at the foot of a cliff hundreds of feet tall. The present town is on the top of the cliff. To get to it you take a cable car—or, if you are more adventurous, and don't mind how you will smell afterward, you ride a donkey up the winding and fairly scary path to the top. Then you look out across the bay.

Down below is the cruise ship you came in on, looking about the size of a bathtub toy. Out in the distance there are a couple of islands guarding the entrance to the harbor. And you realize that those islands, and the seven-hundred-foot cliff you are standing on, are only the fragments that still remain of the crater rim of that immense volcano.

When Santorini's volcano erupted, it caused the mother of all tsunamis—tidal waves—which is very likely what caused the collapse of the previously thriving Cretan empire, a few islands away. How big was that tsunami? Very likely the biggest any human eye has ever witnessed. Some sketchy signs seem to indicate that its crest left debris on mountainsides in Greece and Turkey a quarter of a mile above sea level. Oh, yes. It was *big*.

PERHAPS YOU'VE ALREADY seen Santorini's great crater and the frozen dikes on Etna as well as the tumbled trees and dammed streams around Mount St. Helens and Vesuvius's fossil city of Pompeii. Maybe you're tired of looking at places where something happened once but nothing much is happening right now, and would like to see something that is actually in action for a change.

Well, why not? It happens that there is a reliably performing volcano not far from Etna itself. That one is Stromboli, sometimes called the lighthouse of the Mediterranean for its flaming clockwork eruptions that (it is said) once helped mariners safely navigate past the Aeolian islands off the Sicilian coast.

I'm sorry to say that I've personally seen little of Stromboli. I've actually viewed it only once, and that from the deck of a troop transport long ago and in broad daylight. It's best to see it closer and when the sky is darker. One of these days I hope I may, for, if my friends in the business do not lie to me, that sort of viewing is readily available to anyone who wants to make the trip. When Stromboli is in full form, which is usually, it throws up a fountain of fire every fifteen minutes or so, and guides stand ready to conduct any traveler willing to invest the fee and the fairly hard work involved in climbing up the side of the mountain at night for a ringside seat.

But there are a couple of places that I personally prefer when I wish to see volcanoes in action. They are both islands; and in both cases, although one is a full-fledged independent nation and the other a state of the union, they don't just *have* volcanoes. Volcanoes are essentially what they *are*. These islands are Iceland and Hawaii.

MOST AMERICANS DON'T think of Iceland as a suitable first choice for a vacation. The very name puts people off. It sounds cold and bleak. Even Iceland's biggest fans (of whom I am one) would have to admit that there's a lot of truth to that impression. The island's northern edge just about touches the Arctic Circle.

Still, Iceland's winters are milder than you would think. That great Atlantic Ocean heat pipe, the Gulf Stream, still has a little warmth left to give to Iceland at the far north end of its course, and so Iceland in January rarely gets as cold as, say, Chicago.

Iceland owes its existence to the ocean-spreading mid-Atlantic Ridge, which itself is no more than one small section of the mightiest mountain chain on planet Earth, the forty-thousand-mile-long mid-ocean ridge that girdles our world.

The midocean ridge is not really in the middle of all the oceans it traverses. In its eastern Pacific stretch it presses close against the American coasts. It isn't quite continuous, either. If you follow it east to west, as the Sun moves, it starts just off the coast of Baja California, then strikes out southward across the Pacific Ocean, going south of New Zealand and Tasmania. Then through the Indian Ocean, rounding Africa's Cape of Good Hope at such a distance that it nears Antarctica; thence north through the Atlantic. Here and there in the Atlantic isolated peaks of the ridge break the surface as remote islands—Tristan da Cunha, Ascension, the Azores—until it reaches its most substantial above-water display. That's Iceland. After that the ridge keeps going northward, under the ice of the Arctic Ocean to a point north of Spitzbergen, but it never shows itself above water again.

It is the volcanoes of the midocean ridge that made Iceland, and the island is still growing. Over the last five hundred years a third of all Earth's lava flows have occurred in Iceland, and it has more than two hundred identifiable volcanoes. Some are quiescent, others are very much not. In 1973 the volcano named Helgafell, on the offshore island of Heimaey, came to life with a major eruption that came close to destroying the port of Vestmannaeyjar; and in November 1963 it was a brand-new and unnamed volcano that poked up out of the sea bottom off another of Iceland's shores to create the little island of Surtsey.

Those aren't Iceland's biggest volcanoes, though. One of the largest and most reliable is Hekla, which has erupted once or twice a century over the thousand years Iceland has been inhabited (and, since the latest eruption was in 1947, another could come along at almost any time). Hekla was first observed by Irish monks around the eighth century. They had fled to Iceland to get away from marauding Vikings; when the Vikings followed them to Iceland they went home again, and the stories they carried of Hekla in eruption are said to have formed the basic details of the medieval description of Hell.

The monster of Iceland's flock of volcanoes is the one named Laki.

Laki is a volcano's volcano. It hasn't done much lately, but when it last erupted, in 1783, it emitted nearly six cubic *miles* of lava, and that was a truly serious blast. That lava flow combined with the floods from melted snow and ice to destroy most of Iceland's scraggly farms and thus kill off three-quarters of the colonists' livestock. The human population was not spared. More than ten thousand people—a quarter of the island's population at the time—died of hunger as a result of that disaster.

Well, you won't see anything like that when you visit Iceland—not unless you have inordinately bad luck—but what you can see is another phenomenon that can be counted on to perform for you. It is worth seeing, too, because it is rare in our world; the things only exist in a handful of spots around the globe—Iceland, New Zealand, parts of the American West, and hardly anywhere else. They are the geysers.

Of course, Americans are among the fortunate few who don't have to leave their own country to see geysers; Yellowstone National Park has them in abundance. But Iceland's are special. For one thing, they are the ones that gave the whole class of objects its name. Our European ancestors had not yet visited either America or New Zealand when, in the thirteenth century, the early colonists of Iceland reported a peculiarly intermittent fountain. They named it Gey-sir (the old Icelandic for "rushing forth"). It's still there, near the little village of Haukadalur, and it still functions . . . though not as well as it used to.

Blame that on tourism. When visitors began demanding to see the geyser in operation, local guides sometimes did not choose to wait for the fountain to spurt forth on its regular schedule. They discovered that they could give Gey-sir a sort of enema by throwing debris or oily liquids into it. That worked, and the tourists got to see their spectacle; but, as your doctor will tell you, too much reliance on laxatives has a deleterious effect on one's natural regularity, either for the geyser or the human bowel. Now old Gey-sir is only a shadow of its former self. But there are other geysers still functioning nearby, and they put on a handsome show.

If you are on Iceland you don't have to go as far as Haukadalur to get a taste of what geothermal water is like, though. Indeed, you don't have to leave your hotel at all, for the whole city of Reykjavik is heated by the hot water that is piped from nearby springs. Those steaming springs gave the city its name, in fact; the first settler from Norway, a man named Ingolfr Arnarson, built his home there a thousand years ago, and he named the area Bay of Smokes—in Old Icelandic, Reykjavik.

There's one other spot in Iceland that endears the island to me, although it is no more than a small and undistinguished-looking valley. It is only a couple of dozen feet wide, and its walls less than that tall. It is called Thingvellir, and Icelanders treasure it for its historical associations. It is where the world's oldest parliament, the Icelandic Althing, first held its open-air meetings, beginning around A.D. 950. The Althing is still meeting and legislating today, though now in modern man-made surroundings.

Those early meetings must have been pretty lively. It was the custom for speakers to stand on the cliff at the west side of the valley—which happens to be the easternmost extension of the American tectonic plate—in order to harangue the audiences who stood, on the westernmost extension of the Eurasian plate, in the valley below. The speakers didn't do this because of any interest in plate tectonics. They did it as a safety precaution. The little cliff made it difficult for dissenters in the audience, all of whom carried their swords to the meetings, to get at the speakers so as to settle their differences in a more physical way.

What's important for the science-minded, though, is that the valley of Thingvellir is indeed the precise spot where the great North American plate is pulling away from the even greater Eurasian one. It is that expansion that has made the valley . . . and, on a larger scale, the Atlantic Ocean itself. My private fantasy, as I stand between those walls, is to imagine that, if I had a few strong auto jacks and some stout rods, I could push them an inch or two farther apart and thus speed up the expansion of the Atlantic Ocean a bit.

THERE'S AN EVEN greater rift valley—in fact, it is called the Great Rift Valley—in East Africa. The city of Nairobi, Kenya, is inside the rift valley; it is an immense plain, a hundred miles wide, formed 80 million years ago, when the center subsided and the land to the west rose. It's a wonderful place to visit, with its herds of great game animals—it is an experience to be treasured to sit in a Land Rover while a herd of fifty or a hundred elephants, big ones and babies, strolls placidly past almost close enough to touch. But the Great Rift Valley is much too vast for any tectonic fantasies.

LIKE EARTHQUAKES, VOLCANOES tend to occur at the borders of tectonic plates, and their favorite site is the perimeter of that vast plate that holds most of the Pacific Ocean. The Pacific is surrounded by what is called the Ring of Fire, an immense circle of earthquake-prone regions and volcanoes. The Ring of Fire stretches up the Pacific coast of the Americas—from the mountains of Peru to Mexico's Parícutin and Popocatepetl, as far north as Washington's Mount St. Helens and Alaska's giants—and returns back down along the Asian shore by way of Kamchatka, Japan, and the islands that spawn great eruptions like Krakatoa.

I can't say much about most of these from personal experience. I've seen even the Alaskan volcanoes only from the air, and I've never seen Kamchatka's large collection of regularly erupting ones—famous Klyuchevskaya, the tallest in Russia at 15,584 feet, and all the scores of others—at all. Even Mount Fuji, venerated Fuji-san to the Japanese, I've only managed to see from the window of a high-speed bullet train on the Tokyo–Osaka run. (And actually the last few times I took that train not even then, because the galloping Japanese air pollution hid the mountain from sight.)

But I have left my personal favorite for the last. It isn't in the Ring of Fire. It's situated where you would least expect it, in the very center of that great Pacific plate, and it is the volcano named Kilauea, on the Big Island of Hawaii.

WHY AM I so fond of Kilauea? For one thing, it's simply that it's in beautiful Hawaii, always a treat to visit, and no more than a short drive from my favorite Hawaiian city, Hilo. But Kilauea has one other great advantage for the spectator. Kilauea is a gentle, viewer-friendly volcano. It generally oozes lava and almost never explodes in any really dangerous way, but it is almost always doing something worth watching.

As you drive around the rim of the central crater you pass through a field of sulfur-stinking fumaroles on one side, just where the old Hawaiians used to toss human sacrifices off the cliff to the volcano goddess Pele. Farther along are the remnants of old flows, now hardened, and you can see where they buried buildings and parts of highways. Almost back to the starting point, at the hotel where if you wish you can take a room that looks out over the crater (and where there is also an attractive small museum to visit), is a steep path that allows you to go down and walk on the skin over the permanent lava pool inside. A hundred years or so ago, when Mark Twain visited the place, he had no such opportunity. The rock skin had not yet formed.

The caldera was still filled with hot, liquid lava. Now and then some new vent will open with fountains of fire visible from miles away.

You may not see fountains of fire from there, but for some years now Kilauea has been pretty continuously spilling one large lava flow down to the sea at the south end of the island. When my wife and I were cruising around the Hawaiian islands to watch the 1992 solar eclipse, our ship circled around that southern tip of the island at night. I saw a view of Kilauea's majesty that I had never seen before, and it was spectacular.

You see, when lava streams flow out of an erupting volcano, they cool down from the outside in. After a while what you have is a sort of hollowed-out pipe of solidified rock, with molten lava still flowing inside. If the lava flow stops, all the liquid runs out of the end of the tube, and then, when everything has had time to cool, you have six-foot-diameter lava tubes that you can walk around in. There are several such lava tubes in Volcanoes National Park, and one, the Thurston Lava Tube, is easily accessible.

But when the flow just keeps on flowing, as Kilauea's has been doing for some time now, sometimes the hot lava eats through the hardened shell. Then the molten lava breaks out in patches of bright, orange-red fire.

That's what we saw that night, as our ship rounded the cape and the flow came into view. What it looked like, more than anything else, was the campfires of a great army—maybe of old King Kamehameha's host, as he was waging his long war to conquer all the other kingdoms of the islands—spread out along the hillside, the leakage of lava making bright fiery patches in the darkness. And down at the end of the flow, where the molten lava struck the sea, was the most spectacular fireworks display you can imagine. As the lava hit the water it exploded into rockets and Roman candles, with a sputtering, artillery sound that we could hear a mile offshore.

Spectating science is always rewarding to the mind, but sometimes it produces sights of overwhelming beauty, as well. In my experience few have ever been more so than that supernaturally brilliant spectacle as the world's interior magma met and battled against the sea.

THAT BEAUTY, THOUGH, conceals what may be a pretty worrisome danger.

As all that molten rock hits the water and cools, it adds new territory to the southern end of the island of Hawaii—about four inches a year, as measured in 1994 by the space shuttle *Endeavor*. Some seismologists fear that it is adding too much, too rapidly. The worry is that these millions of tons of added rock may overbalance their roots on the steep underwater slope of the island. If that happens, all that ponderous mass may crack loose and slip to the bottom of the sea. . . .

And if that happens, the splash of that landslip will produce a tsunami. Not a small one, either. Very possibly the largest the human race has ever known—even larger than the one from the island of Santorini (or Atlantis) more than three thousand years ago—and one that could race across the Pacific and strike the shores of California and Japan and a dozen other Pacific coasts, with results that hardly bear thinking of.

Is that actually going to happen? No one knows for sure, but some big slides of that type are known to have happened in the past, though none in historic times—but then there was no one on Hawaii to write history until quite recently. And there are troublesome signs. In fact, there is a visible narrow crack that is spreading along the surface of the southern part of the island, Hilina Pali, just where the split is most likely to occur. Seismologists keep their eyes on that crack to see if it widens. (And so, as opportunity permits, do I.)

THERE'S ANOTHER IMPORTANT Hawaiian volcano on the island of Maui, the next one over from Hawaii itself. That's Haleakala, with a crater that is accessible by foot for the dedicated hiker or by means of a six-hour mule ride for everyone else. It's not entirely dead, either. Haleakala erupted as recently as 1790 and may well give us another farewell appearance or two before quitting the stage for good.

Yet another Hawaiian volcano is in some ways the most interesting of them all, in spite of the fact that you have no chance of actually seeing it.

That other volcano is called Loihi. The reason it is invisible to the visitor is that its peak is just about a mile under the surface of the Pacific Ocean.

To understand what Loihi is all about, we need to think about how the Hawaiian chain was formed. What creates the Hawaiian volcanoes—what has formed the entire chain of the Hawaiian islands—is a "hot spot" under the bottom of the Pacific Ocean. There are other such hot spots in the world, though why they exist no one knows. What scientists do know is that at such points the crust appears to be particularly thin, and so there is a nearly constant outflow of lava from the magma under the crust. And this particular hot spot has remained in the same place at least 70 million years, while the Pacific plate has slowly moved over it, traveling some fifteen hundred miles toward the north and west.

The result is that this hot spot has produced, one by one, a series of giant volcanoes thrusting through the Pacific tectonic plate. Each one starts at the sea bottom and grows until it reaches the surface and becomes a volcanic island. The string of these volcanoes has, in fact, become the entire chain of the Hawaiian archipelago. Each began life as a single volcano or a cluster of nearby volcanoes caused by the hot spot under the sea bottom. Then, as the motion of the Pacific plate carried it away from the hot spot, it remained as a dying volcano, its influx of new material dwindling and finally stopping entirely.

Altogether there have been about 131 such events in the past 70 million years. The oldest volcanic cones have long since disappeared; they have been eroded away by the action of wind and water, leaving only seamounts (undersea mountains that do not reach the surface). For the ones slightly younger what remains are shoals and tiny islets—Midway Island is one. But then you come to Kauai, the oldest of the major islands of Hawaii; although erosion has been working on it for a few million years, a good deal of it is still left. Moving toward the south and east you pass Oahu and Maui, each one younger than the one before, until you come to the newest of all, the Big Island of Hawaii itself.

The Big Island is still close enough to this great underwater hot spot to be receiving outflows of magma—thus Kilauea. But the Pacific plate hasn't stopped moving. The hot spot below already has begun to make a new volcano, a couple dozen miles southward of the Big Island.

That is the underwater volcano called Loihi. Scientists from the Hawaiian Underwater Research Institute have been poking around it in a submarine, and they confirm that it is growing.

Loihi is now two miles high above the sea bottom, with a half mile of water still above it. Sooner or later it will add that half mile to its height. Then it, too, will break the surface and begin to become an island on its own. Come back in about fifty thousand years and see.

Chapter 6
WATER, WATER EVERYWHERE

Dams, Locks, Tsunamis

We've given a lot of thought to the science-spectating potential of the land, but perhaps we're shortchanging ourselves. Our planet's surface, after all, is predominantly water. Now let's turn our attention there.

Water is wonderful stuff to have around. You can drink it, you can bathe in it, you can float ships on it, you can (if you're careful) dump your wastes in it and let it carry them away. We couldn't live without water. In fact, we *wouldn't* live without it. If there were no water for us we would never have been born, since more than three-quarters of our bodies, and of the bodies of all our evolutionary ancestors, is nothing but water.

But water has its mean side, as well. If you want to see what kind of punishment too much water can inflict, there is no better place to begin than a visit to the pretty little bayside park in the city of Hilo, on the Big Island of Hawaii.

We're talking tsunami here. That is, we're discussing the things that are sometimes miscalled tidal waves, though they have nothing to do with tides. Hilo knows all about tsunamis. That's a matter of geography, because the city sits on the edge of a shallow, funnel-shaped bay on the western coast of the Big Island. In 1946 there was an earthquake in the Aleutian Islands, far away in the Bering Strait, and the tsunami it produced arrived in Hilo on April Fool's Day. A dozen successive waves roared up over the business district of the city. As the waves washed back to sea, they took the business district with them. Buildings, desks, streetlights, cars, typewriters, and all are now embedded in the Pacific Ocean's sea-bottom sludge, and 139 people died.

When Hilo pulled itself together and rebuilt after the disaster, the residents prudently relocated their new business district farther from the bayside . . . and on higher ground. That left them with the planed-flat area swept by the tsunami.

There they decided not to rebuild. Instead, they converted it into a pretty little waterside park. It's a pleasant place to take a picnic basket, and when you're done with your lunch you can pay a visit to the small museum in the park to see just what the tsunami did to the city of Hilo on that spring day in 1946.

Oh, and before you leave Hilo, one other thing. Look across the bay and you'll see a ring of hotels just above the water. Since this is where tsunamis are known to have blasted inland, and surely will some day again, you may wonder what kind of crazy people build expensive hotels right there.

Well, maybe not entirely crazy. If you drive around the bay to Banyan Drive, where the hotels' entrances are, you will notice that you need to go up a ramp to get to the hotel lobby and restaurants, which are up on the second floor. What is the ground floor for? Why, it's a car park. It has sturdy pillars to support the weight of the floors above, but it has no solid walls.

See how it's supposed to work? If one of those ten-foot tsunami waves comes along, it's meant to roar right through that lower level, leaving the important parts of the hotels overhead untouched. The hotels wouldn't even have to evacuate their guests. They could stay right in their upper-story rooms, gazing down at the harmlessly raging waters below, with their cameras going to show the folks back home.

That's good planning, turning a hazard into a tourist attraction, isn't it? Of course it is . . . well, it is unless the planned ten-foot tsunami turns out to run fifteen or twenty feet instead.

AN EARTHQUAKE IS one of the best ways to cause a tsunami, but other things can do it, as well. A volcanic eruption under the sea, maybe even the collapse of a high seaside cliff or the impact of a really big meteorite, anything that suddenly displaces a large volume of water can send a tsunami across the sea. Water is pretty incompressible. There's no elastic give; that displaced water simply has to go somewhere else. It becomes a wave, which moves outward from the original disturbance, like the ripples when you throw a stone into a pond. And that tsunami wave moves along at jet-plane speed, so rapidly that it can cross an entire ocean in hours.

How big is this wave? That partly depends on the depth of the water under it. In midocean, even a large tsunami may amount to only a matter of inches at the surface; ships sailing over it aren't likely even to notice it's there. But as it approaches a coast and the water becomes shallower, the same volume of water has to move. So the height of the wave increases. By the time it strikes a shore it isn't measured in inches anymore. It can be twenty feet or more of water rolling up onto the land, and it can carry away whatever it touches.

Tsunamis are killers. In Alaska, the Good Friday earthquake of March 27, 1964, produced a tsunami that swept twenty-six people to their deaths off a pier in Valdez, taking the pier along with them.

Alaska and Hawaii are the most likely places for tsumanis in the United States. Hilo gets hit every eight or nine years on average, though most of its tsunamis are too small to be worrisome. But a really big one can hit, because thousands of years ago, before even the first Polynesians had discovered the islands, there appear to have been tsunamis that were truly immense. Coral remnants have been found 150 feet above the sea on the island of Molokai and a whopping 1,000 feet up on Lanai; only a tsunami could have put them there.

In terms of tsunami frequency, Japan is the world champion, with the whole Pacific Ocean's Ring of Fire to produce earth movements. All through Japan's long history, recurring tsunamis have repeatedly destroyed or damaged its seaside cities and killed many thousands of its people.

It isn't just the Pacific Ocean that throws a catastrophic tsunami onto the land every now and then. The one that struck Lisbon, Portugal, in 1755 probably took more lives than any other in recorded history. Before the big wave arrived, the water receded from Lisbon's harbor, leaving the sea bottom bare. Curious citizens went out onto the exposed sea bottom, and when the giant wave came in they all died. So did a good many who had never left the shore.

Earlier than that, before any histories were written, there were far greater tsunamis in the Atlantic. One such was seven thousand years ago, when a great underwater landslide occurred between what are now Scotland and Norway. The tsunami that resulted swept up over the coasts of both places and all the islands in between. Archeologists believe that Stone Age tribes inhabited many of those coasts, but, with no written history, there is no way to be sure. They would have been obliterated.

And the greatest tsunami of all, at least the greatest that left any traces, was the one that struck off the coast of Mexico's Yucatán

Peninsula 65 million years ago. That's the one that is supposed to have darkened the whole planet with the debris it threw into the atmosphere, killing off the dinosaurs and most other living things on Earth.

That one did leave recognizable signs. Some of them are a long way from the impact point, and they're not easy to find. But even after 65 million years they are definitely still there—deposits of sand and gravel thrown up into the Maya Mountains in the little Central American country of Belize, sand layers two feet thick in Texas, and traces of that enormous wave as far from its source as South Dakota.

TSUNAMIS AREN'T TIDAL waves, but real tidal waves do exist. Just as Hilo's funnel-shaped bay concentrated the tsunami that struck it in 1946, there are other bays that can turn even the normal tides of Earth's oceans into giants, dozens of feet high. One such is Nova Scotia's Bay of Fundy, and one of the regrets of my life is that the only time I saw it for myself I couldn't spare the time to wait and watch that huge tide roll in.

Since tides of this sort happen on a daily basis, they are expected and allowed for; they don't kill, and they don't destroy property. But there is another mechanism by which water marches up onto the land, and that one does kill. That's the storm surge.

The storm surge is deadliest when it comes with a major hurricane. The atmospheric pressure in a hurricane is so low that it actually lifts up the level of the sea beneath it. Then, when the hurricane moves onto land, it brings that higher water right along.

For the Lower 48 states, storm surges are a lot more common than tsunamis. Hardly any part of the Atlantic and Gulf coasts of North America, from Long Island, New York, to Corpus Christi, Texas, has escaped being battered by one of those over the years. But for sheer catastrophic damage, no storm surges in North America can match the disasters of Bangladesh. Much of the country lies only a couple of dozen feet above normal tides. When a hurricane (locally they are called typhoons, but they're the same thing) whips up in the Bay of Bengal, the tides become extremely abnormal from the storm surges. Then thousands of square miles of Bangladesh can be flooded, and the loss of life has been in the tens of thousands. In 1970 it was more than that. Half a million people died in one of the worst natural disasters ever.

* * *

NO SENSIBLE PERSON really likes being in the middle of a hurricane, so storm surges are not attractive spectator sports. Still, there is much in the oceans that is worth looking at, especially for the science-minded.

For one reason or another I've never been able to do scuba diving, but those who have are rapturous about what you can see when you swim with the fishes. Never mind. I've been able to see enough to delight me from a civilian submarine in Vancouver, British Columbia, from glass-bottomed boats in Bermuda and the Seychelles, or from snorkeling in the peaceful lagoon of the South Sea island of Mooréa. My wife tried snorkeling for the first time at Australia's Great Barrier Reef; it made her an instant convert, and since then her favorite travel destinations take us to places where the water is clear, the temperature is warm, and there are fish, corals, and bottom-dwelling creatures to observe.

Coral isn't just pretty; it is alive. It is made up of the skeletons of tiny invertebrate animals called Anthozoa, of which there are thousands of different species. As free-floating mites they attach themselves to the first hard substance they bump into. Sometimes that is rock; more often it is coral that is already established, and there they proliferate.

The corals have evolved to withstand most of nature's challenges, but not man's. In spite of their hard external skeletons, corals are delicate. A ship's anchor can kill a patch of coral, a freighter running aground can damage an entire reef. When wooded islands are over-logged, their ground cover is destroyed, their streams carry vast amounts of mud into the sea, and the mud chokes the corals. In recent years a good deal of coral has been killed by "bleaching"—the living parts of the animals die, leaving the skeletons pale and dead—and it appears that this may be due simply to small rises in the water temperature.

But when they are left on their own, they can produce such marvels as the Great Barrier Reef, more than a thousand miles long. To get there you fly to any of the towns along Australia's east coast, and then you take a boat some miles offshore to the reef itself. Suddenly you are floating in the shallows again, with a few hummocks here and there actually above water. That's when the snorkeling begins.

It isn't just human enemies that despoil corals. On the Great Barrier Reef the worst natural threat comes from the Crown of Thorns starfish, ugly things with more limbs than you might think a starfish should have. They eat the living parts out of corals, leaving those parts of the reef dead, and the starfish themselves are extraordinarily hard to kill. If you stab them, they heal; if you cut them into pieces, each piece regenerates to become a new adult. What they cannot survive is being dried out, and so many Great Barrier Reef divers spend a lot of their time pulling the things out of the water and tossing them onto one of the above-water hummocks to perish.

Corals can't survive in deep water, which is surprising when you consider that some coral reefs are dozens or even hundreds of feet deep. What happens is that many tropical islands, like those of Hawaii, begin as volcanoes. They thrust up out of the sea; then the volcano dies, and the island begins to settle. The corals continue to grow in layers as the water deepens.

Often the corals form a nearly circular reef around the island, creating a lagoon between reef and shore. In my personal opinion, such lagoons are the best places this world has to offer for gentle, relaxed swimming. The reef keeps such nasties as sharks outside, the water is warm, and there are no waves.

And, of course, the water is salt, which makes swimming less strenuous; salt water is denser than fresh, and thus more buoyant. Where the salt content is extreme, as in Israel's Dead Sea, it is nearly impossible to sink. Freshwater swimming requires more energy simply to stay afloat. (For which reason a few competitive swimmers have tried to figure out ways to become more buoyant—going as far, as the East German swim team did in the 1976 Olympics, as to pump air into their, excuse the expression, rectums to make them lighter.)

The other wonderful thing about most tropical lagoons is that their waters are so beautifully clear. Scientists measure the clarity of water with a thing called a Secchi disk, about the size of a pie plate, divided into quadrants of alternating black and white. You lower the disk into the water and measure the point at which the difference between the quadrants disappears. In pristine water it is still visible fifty feet down. The Arctic Ocean does sixty feet or better; the people of the Cayman Islands claim a hundred. And in New York Bay, on a good day: maybe four.

Water clarity, though, is not always a sign of good health. Since the invasion of the zebra mussels in the 1980s, parts of the Great Lakes have cleared amazingly. This is because the mussels are voracious filter feeders, eating all the algae that otherwise would cloud the water. There are other creatures that would eat that algae if they could, but the zebra mussels outconsume them so that the others die. And so that water is beautifully transparent, but it is also dead.

YOU DON'T HAVE to go to Australia or the South Pacific to see a coral reef. There's a big one off the coast of Belize and a much smaller, but much handier, one off Florida's chain of keys; that one is protected as America's only underwater national park. The diving is fine, but if you go you should remember the Golden Rule of visiting any park: "Take nothing but pictures, leave nothing but footprints." Here, of course, you don't leave footprints, but it's even more important to resist the temptation to take anything. The other way a coral reef can be destroyed is by divers breaking off the prettiest bits and carrying them away.

Coral reefs aren't just pretty. They also protect shorelines by breaking up some of the most threatening aspects of storms, as do chains of barrier islands.

America's Atlantic coast has a fine chain of barrier islands, all the way from New Jersey to Florida; their existence explains why it is possible to take a small boat all the way down the coast without venturing into open ocean.

Next time you are in Sea Bright, New Jersey, or Sea Island, Georgia, or any of the little islands off the coasts of the Carolinas, notice that they are pure sand. They are not much more than sandy beaches that have been isolated from the mainland, and it is the nature of sandy beaches to be temporary. They are built up naturally by the action of waves, tides, and storms. They also are naturally destroyed in the same way. Left to itself, each one of these islands is constantly changing its shape, growing and diminishing, opening channels to the Atlantic and closing them again.

The trouble is that they aren't left to themselves. For the past couple of centuries people have been building on them, and for the same period of time the things that were built have been frequently destroyed when storms come and the islands change their shape. Flourishing little towns like Sea Bright do their best to prevent such change and destruction. They build ten-foot-high seawalls. They try to preserve their beaches on the far side of the walls by dumping thousands of tons of rock to build jetties extending out into the ocean and interrupting the currents that flow along the shores. But sand is only sand, and the ocean never gives up. A barrier island may survive through a thousand violent storms, but then along comes the thousand and first and it is gone.

THOSE LITTLE COASTAL currents that move sand from one beach to another are only the smallest of the immense oceanic rivers that flow all over Earth's seas. Most are invisible to the naked eye—but not all.

The most famous of those great oceanic currents is the Gulf Stream. If you happen to be in a high-rise hotel room facing the ocean in Miami Beach, you can look out your window and observe a line of indigo water, far out. That's the stream. It is by far the world's largest river—flowing through the ocean, rather than on land—and the warmth it transports from the Gulf of Mexico to the European coast is what keeps Iceland, the British Isles, and much of northern coastal Europe warm enough to support civilized life.

The stream flows northward, but of course all that water cannot continually be removed from southern waters without a southward return flow to replace it. What pulls the stream northward is downwelling cold water in the Arctic, much of it spawned when surface waters are frozen into ice.

Salt water does not freeze, even at Arctic temperatures. The part that freezes out into ice is fresh water, and so the water that is left behind is saltier than before. Thus it is denser; thus it sinks to the bottom, and ultimately slides southward, well under the surface, as the returning flow, the other half of the Gulf Stream.

However, there is a worry here. Less ice is forming at the northern latitudes than in recent centuries, perhaps because of global warming. If those downwelling waters stop and that return flow is impaired, the Gulf Stream itself may diminish. It may even stop, as it is thought to have done on previous occasions in the geologic past. And then the climate of England would become more like that of other areas at the same latitude, such as Labrador.

SO FAR I'VE been talking about the salt seas of Earth. But it is freshwater that we depend on for life, and freshwater, too, has its mighty streams—and can do its mighty damage.

If, sometime, you are in the neighborhood of Lawrence, Kansas, drive to the nearby intersection of Routes 24 and 59. You'll find yourself in the middle of a broad valley. Where the roads meet there is a curious concrete structure, shaped like a teepee—built as a roadside nightspot, later a sculptor's studio. No water is in sight. The nearest bend of the Kansas River is a half mile away.

But in 1951 the Kansas River overflowed its banks. A marker high up on the side of the concrete teepee, twelve feet above ground, shows how deep the water was at that point as the flood filled that valley from horizon to horizon, and the streets of North Lawrence were traversed in rowboats.

That was only a relatively little flood. There have been many that were far greater, and the greatest of all in recorded American history happened as recently as 1993.

It began with a storm on March 14. It was a big storm, and it didn't stop. It continued for four days, and in that time it dropped 50 million acre-feet of water onto the Midwest—that is, in each of those days, as much water as goes over Niagara Falls in a month.

Then there were more storms. For the states where the Missouri and Mississippi rivers begin—Illinois, Iowa, Minnesota, Montana, and the Dakotas—it was the wettest summer on record. Over a period of six weeks as much water fell on those areas as there is in Lake Erie. So the rivers began to rise, as much as forty feet at St. Louis, with water reaching almost to the foundations of the great Gateway Arch. First to last the rivers overtopped or breached more than a thousand levees, flooding 20 million acres of previously dry land, causing forty-eight deaths and leaving seventy thousand people homeless. For five hundred miles along the river bridges were damaged and closed to traffic; more than a score of airports were flooded; and transportation came to a halt. People who had lived on one side of the river and commuted daily to work on the other commuted no longer.

As late as July, when the bridges were open again, I happened to drive across the Missouri River at Kansas City. It was a road I knew well, but it looked strange. Areas on either side of the road that had been open meadow were now lakes, and they stayed that way for months.

The 1993 Mississippi flood was an expensive event, and not just in dollars. Because so many croplands were flooded, America's corn production dropped by a third from the previous year. Pastures were scoured clear of grass and replanted by the flood with waterborne seeds of such weeds as cockleburr and pigweed—not a benefit for the dairy farmers, because those plants are toxic to cattle. The flood flushed the zebra mussel out of the Great Lakes, where it had damaged or blocked water works and power plant intakes, into the Mississippi system, where it began to do the same.

THOUGH THAT 1993 flood was the most damaging in American history, there have been plenty of others. Floods are nothing new along the Mississippi. When Hernando de Soto's men first reached the river in 1539, it happened to be in flood, and they were astonished to find they couldn't see its far shore—not surprising when you know the river's history; sometimes it has spread as much as a hundred miles. The Indians who lived there at the time had found a way to deal with the problem. Their villages were all located on high ground; when there didn't happen to be any high ground, they made some. Laboriously they dug and carried earth to make a hillock to settle on, and when the river flooded they simply retreated to the villages until the water went down.

When Europeans settled the area they were not that smart. In 1737 the river rose around the new French trading post at Nouvelle Orleans. To protect it against the next flood they raised the first of the famous New Orleans levees.

That one was only a few feet high. The levees have been growing ever since, in a continuing effort to keep the river out of New Orleans.

That's what the levee does. It is also a first-rate tourist attraction. If you are visiting the Big Easy you will of course go to the levee on a Sunday morning for your breakfast of French coffee and the fried pastries called beignets. When you've finished eating you probably will work off the calories by strolling along the levee—perhaps to visit New Orleans's fine aquarium or the levee-top shopping mall, or just to watch the ships go by.

New Orleans is a wet city. If you visit its cemeteries you will see that no one is actually buried there; since the water table is just below the surface, interments are in aboveground mausoleums. The levee keeps the city from being a great deal wetter still; it is the levee that keeps New Orleans business district, hotels, and French Quarter from being a shallow lake. That's the good part of the levee, but levees have a drawback, as well.

The trouble with levees is that every one you put up increases the risk of flooding for everyone nearby. So in self-defense, your neighbors put up levees of their own and then their neighbors do the same. The result is today's Mississippi, with the riverside diked almost uninterruptedly from St. Louis to the Delta. So now the corseted river runs straighter and faster, as much as twenty-five miles an hour in some stretches. And when it does flood, and a levee fails, the pent-up floods scour buildings and topsoil away—in 1993 such a flood even exhumed bodies from cemeteries by the thousand.

Most of the damage from that 1993 flood has been cleared up, but if you want to see something of what it was like, go to Fort Madison, Iowa. There, on the banks of the river, the old Santa Fe railway depot has been turned into a flood museum. In it they have preserved Mississippi mud, sandbags, and ruined appliances. Looped videos of the televised flood warnings and newscasts are shown, and there are (of course!) souvenirs to buy . . . including some punning T-shirts that, rather wittily if not all that tastefully, say:

The Flood of the Century
Row vs. Wade
It's Your Choice

The Mississippi is not the only river that is penned in by endless levees. In Europe the Rhine is almost as heavily diked—or "rectified," as Johann Tulla, the engineer who began the process in 1817, preferred to say. Before Tulla much of the Rhine meandered in many streams over a great floodplain; it was said that the border between France and Germany changed every time there was a heavy rain. Tamed and straightened, the Rhine became the great waterway for transport that helped Germany through the Industrial Revolution. The cost was great, though. Pollution and industrial waste killed the Rhine. Although there have been strenuous attempts to clean it up in recent years, the Dutch, through whose country the Rhine flows to the sea, still have to dredge out and bury thousands of tons of toxic sediment every year.

For a different kind of technological fix for flooding, consider London.

London is nowhere near as vulnerable as the cities along the Mississippi. It is low-lying, though, and in recent years its Thames River has been subject to unusual storm surges and increasingly high tides that come all the way in from the ocean, possibly for reasons related to global warming.

Therefore, Londoners have built the high-tech system called the Thames Barrier. If you're in Greenwich to visit the old astronomical observatory and the *Cutty Sark*, look downstream to the river at the neighboring community of Woolwich. There may be little to see, because most of the barrier is retracted underwater when the Thames is quiescent. But when the river threatens to flood, giant steel clamshells are rotated up from the bottom, making a sort of temporary dam and preventing the flood from reaching the city.

Is London now safe from flooding? At least until the year 2050, the engineers believe. But if the global warming proceeds as some fear, after that time even the Thames Barrier may be unable to keep London dry.

WHAT MAKES FLOODING rivers devastating is that people insist on building and living on natural floodplains. Without such encroachment, the rivers would simply flood harmlessly from time to time and then go quietly back within their normal banks.

Indeed, floods can produce great benefits. As far back as Cleopatra's time, Egyptians counted on the annual flooding of the Nile to refresh the country's farmlands; each flood spread new layers of richly fertile mud every year, which is the principal reason why Egypt was Rome's breadbasket in the days of the empire. And, indeed, the annual floods continued to help Egyptian farmers for many centuries after that, until the great dam at Aswan put a stop to the floods.

Like levees, dams sometimes fail, and when that happens the results are spectacular. The failure of a relatively small dam produced the famous Johnstown flood in America; fortunately for the many millions of people who live downstream of dams in the United States, there hasn't yet been a total failure of a large one.

Or, at least, not of one that was built by human beings. The greatest dam failure known happened a long time ago. That was when Lake Missoula broke through its restraints and scoured the Pacific Northwest.

Perhaps you've never heard of Lake Missoula. That's not surprising, because no human being ever saw it, or ever will. Lake Missoula doesn't exist any longer. It was formed when the North American glaciers began to retreat at the end of the last ice age; meltwater formed a giant lake held in place by the ice that surrounded it.

That was what scientists have termed Lake Missoula. In its prime, fifteen thousand years ago, it was a true Great Lake—not as big as Lake Superior or Lake Michigan, but bigger than any of the other three, and, at nearly half a mile, deeper than almost any other lake in the world. When its ice dam could hold it no longer, the lake burst out and five hundred cubic miles of water came roaring down onto what is now the eastern half of the state of Washington.

A hundred and fifty centuries later, the traces still can be seen. So much topsoil was carried away by the flood that there are some thirteen thousand square miles that cannot be farmed even today. And then there are the scablands.

You've probably seen the ripple marks on a beach or streambed left by the water. That's what the scablands are, but so much water poured out of Lake Missoula that these ripples are *big*. The ridges run fifty feet high. To get a good look at them you need to be in an airplane.

There's one other known set of scablands as huge as those, but the only way you and I are going to see them is in a photograph taken with the help of a spacecraft-mounted telescope. These scablands are the remaining evidence of some truly enormous long-ago flood, in a place that now seems to have no water at all. That place is in the deep valley called the Ares Vallis, and it is on the planet Mars.

FEW MAN-MADE DAMS impound as much water as Lake Missoula, but there are a lot of them. Collectively they hold a lot of water, from giants like Lake Nasser, behind the Aswan Dam, or Lake Mead, backed up behind the Hoover, to the multitude of tiny ones that impound water for turning a mill or running a small domestic electricity generator.

How much water is "a lot"?

Well, it is enough water that if all those dams didn't exist, and all that water had been allowed to continue to flow to the sea, the average sea level around the world would be a little more than an inch higher than it is now. Astonishingly, it is even enough water to affect the length of the day. Most of those dams are in the temperate regions of the Earth, so that that great weight of water is held some distance farther from the equator than it would be without human interference.

You've seen how an ice skater will pull her arms close around her body to spin faster? That's the effect of what is called the conservation of angular momentum, and it's the same with Earth. Because those millions upon millions of tons of water are now held a bit closer to the central axis on which Earth turns, they have sped Earth's spin up and shortened our day by an extremely small, but measurable, fraction of a second.

Dams are beautiful things to look at. If you happen to be visiting the Grand Canyon, it would not be much out of the way to drive downstream to the Hoover Dam. Or, if in Redding, California, to take the short drive to see the Shasta Dam and beautiful Mount Shasta nearby. As you drive, or stroll, across the top of one of these colossal structures, you can't help thinking that the human race is capable of remarkable works. The lakes the dams impound are scenically spectacular. And the work they do makes many of our lives easier and even permits us to live where we choose. Many great cities could not survive if engineers hadn't dammed up rivers or streams to preserve the water that fills their water systems. I got some idea of the magnitude of these works years ago when I had a summer home in Ashokan, New York, looking down on the great reservoir whose waters were piped scores of miles to serve New York City. And the hydroelectric power from some of our great dams provides a significant fraction of the world's energy needs—and does it without burning polluting fossil fuel. In the past, the mechanical power from relatively small dams turned the waterwheels that began the Industrial Revolution. Today dams not only provide us with electricity, drinking water, and irrigation for our farms but with recreational facilities, scenic beauty, and protection from some kinds of devastating floods.

They don't do all of this, however, without some heavy costs.

The scenic beauty of Lake Mead conceals the now-lost scenic beauty of the Colorado River valleys that were flooded to make the lake. The dollar cost of new dams is staggering—one reason why they are one of the favorite pork barrels for members of Congress busily diverting tax money to their own districts. Dams are at risk from earthquakes, and sometimes the sheer weight of the water impounded *causes* earthquakes—so far no major ones, but the jury is still out. Dams interfere with salmon runs, so that many rivers that used to have vast quantities of salmon now have hardly any; a case in point is the Columbia—Snake River system that goes as far inland as Idaho and has more than a hundred dams to handicap the salmon in their rush upstream to breed.

Efforts are made to ease the problem with "fish ladders." If you go to the Bonneville Dam tourist center, you can watch the salmon jump up one of those ladders through underwater windows. Many are saved there, but most salmon are not that lucky. And, inevitably, in the long run dams silt up. All the mud and rock fragments a rushing river carries along with it come to the still water behind a dam and drop out, so that the bottom fills and the amount of water held is reduced. Over time, when their basins have filled, most dams will cease to operate. And the most heavily dammed rivers lose so much of their water to dams and diversions that there isn't much left of the river.

The Colorado River is the textbook case of this. When it reaches the Mexican border, there's so little flow remaining in the river that it's a serious cause of diplomatic friction with the Mexicans, who would like to have more of that water for their own purposes. By the time it reaches the coast so much has been diverted that the Colorado doesn't have a mouth any more. The sludgy, squishy remnant of its once-rushing flow loses itself in marshland before it can reach the sea.

So dams definitely have their downside. Wouldn't it be wonderful if we could get some of the benefits of dams without the worst of the costs?

Actually, sometimes we can. As it happens, the Chinese figured that one out long ago.

IF YOU CHANCE to be in the Chinese city of Chengdu, it's worth the trip to visit the old Dujingyan Irrigation System.

To get there you take the main road that leads west from Chengdu, which is also the road that leads to Tibet. Since just about everything the Chinese need in Tibet has to be trucked in from outside, it's a well-paved six-lane highway. (That is, it's usually six lanes. Not during harvest time, though. Then the local farmers find the hard surface of the Tibet highway perfect for spreading out their winter wheat and rapeseed to dry—and at the same time to be threshed automatically by the tires of the trucks driving over them.) And a few hours out of Chengdu you come to Dujingyan.

The Dujingyan waterworks is not a marvel of twentieth-century technology. It's a lot older than that—2,200 years older, to be exact. The installation has been expanded and repaired a bit since then, but not much. After all those centuries it is still working just fine.

Dujingyan has no dam at all. Instead of a dam, the ancient Chinese put a man-made island in a bend of the river, just upstream from where they wanted to divert water for irrigation. The artificial island splits the river's flow into two parts. One part carries the river down its original course. The other enters a canal that leads to a large area of previously dry farmland.

In order to get the water to these thirsty farms, the builders had to cut a small mountain in half. As they hadn't yet invented industrial-strength explosives to break the rock up, they had to deal with it in a different way. They built fires on exposed rock surfaces, then doused them with cold river water. The rock fractured. They pried it apart with crowbars, then broke it up with sledgehammers and carried the pieces away in wicker baskets.

That sounds pretty primitive, which it is, but all the same there's some wonderful engineering to admire in Dujingyan. First, because there isn't any dam, there isn't any impounded water behind it to silt up and make problems. Second, the artificial island is sited so that most of the suspended sand that comes down at it from the river is carried to the inner curve of the stream, and thus right on downriver, by natural hydrodynamic forces. Thus the irrigation water is left pretty sediment-free. Third, a pretty arrangement of weirs and bottlenecks insures that the farms always get what they need. When the river is low, more of its flow is automatically diverted to the farms; when it is high, the allocation to the farms holds steady and more goes downstream.

And this whole Dujingyan system has been working pretty reliably ever since the time of the Egyptian pharaohs. You have to admire those early Chinese . . . and they didn't even have any computers to help them with their designs.

NOT FAR AWAY, in what is now also China but wasn't when they were built, are some other ancient waterworks that are still in use.

If you drive from the city of Urumchi in far western China down along the old Silk Road to the pretty little oasis at Turfan, you pass through a section of the Gobi Desert. (Actually we should just say "the Gobi"; the word "gobi" *means* "desert.") The Gobi is unlike any other desert I've ever seen. It is made of pebbles rather than sand; the winds are so fierce and the ground so dry that all the finer material has long since blown away, much of it becoming the rich loess soil of central China. But it is as dry as any other. There are occasional oases like Turfan, where once the silk merchants gratefully stopped to rest and rehydrate and now the occupants raise grapes and melons with their springs. The rest is unrelievedly arid.

Turfan's springs, however, do not come from nowhere. They are the end point of underground flows that rise in the Flaming Mountains, scores of miles away. More than a thousand years ago the local inhabitants—genetically they are Turkic rather than Han Chinese—took note of these springs and decided that what nature could do, they could do, as well. So they constructed a network of buried aqueducts from mountains to desert.

As you look from the road over the waterless waste of the Gobi you can see occasional cairns of rock. These are manholes to the buried aqueducts below. For all those generations men have climbed down into the aqueducts to keep them cleaned out and repaired, and the water still flows there today.

It isn't just the Gobi that has underground aqueducts of this sort. The same Turkic peoples built the same sort of systems in many parts of the Middle East, but how long they will continue to flow is debatable. They do need continuing maintenance to operate, and in each generation there are fewer and fewer people willing to climb down and do the hard and hazardous physical work required.

THE ASIANS WERE not the only great engineers of classical times. In Europe the Romans were pretty good at putting water where they wanted it. In Rome itself they were too smart to take their municipal water supplies from the Tiber River, which was generally pretty polluted with waste products and (because one kind of execution involved throwing the deceased into the river) the occasional corpse.

So the Romans built their famous aqueducts, bringing clean water from mountain streams and springs right into the heart of their cities. They built them cleverly—the angle of flow just enough to keep the water coming, in tunnels through hills and on bridges over valleys. And they built them to last. Almost everywhere in southern Europe and parts of Asia Minor their remains still stand, two thousand years later.

Like most things, the Roman water systems were not an unalloyed blessing. When the water arrived in the cities, it was healthful and clean. But when the ancient Roman plumbers laid pipes to bring it to fountains, palaces, and cisterns all over the city, they rolled their pipes out of sheet lead. Lead is easily worked, but it is also slowly poisonous in drinking water; one speculation says that the decline of the western Roman Empire was helped along by the fact that so many leading citizens suffered from the delusions and decrepitude brought on by lead poisoning.

After the Roman Empire split in two, the Eastern Empire continued the tradition of vast waterworks. The capital city of Istanbul—it was still called Constantinople in those days—was laid out with defense against enemy armies in mind. Surrounded on two sides by the waters of the Bosphorus and the Golden Horn, and with the remainder of its perimeter heavily fortified, Constantinople reckoned itself well prepared for invaders. In the event of a siege the people would simply hole up inside the city and wait until the attackers gave up and went away.

Of course, while they were waiting they would need water to drink. Those early Holy Romans knew what to do about that. They constructed a reservoir. Since there wasn't much aboveground space in the cramped old city, they put it underground.

It's still there, right under some of modern Istanbul's buildings, and you can visit it any day of the week. You buy your ticket and walk down a flight of stairs, and then you're in a chamber about the size of a convention hall, though not as high. Pillars hold up the roof, twenty or thirty feet above your head. It's dank and chilly there, whatever the weather is outside. There is a foot or two of dark water below you as you stroll through the place on a catwalk. There are said to be fish there, but I never saw any.

Of course, now it is only a tourist attraction, and just enough water is kept there to show what its purpose was. In the days when Constantinople was fighting its wars, the chamber was kept filled top to bottom. In times of war the city might run out of weapons or soldiers or will . . . but not out of water to drink.

THE GREAT WATERWORKS of antiquity weren't copied by the Western European nations that succeeded them. It was more then a good many centuries before those countries began moving water around on a large scale. They did it for a different purpose, and they did it in a different way. They built canals.

As I said earlier, it was waterpower that turned the wheels that began the Industrial Revolution, but water also played another important part. In the shape of canals, it provided a way of moving large quantities of raw materials to the factories and finished goods out.

First to last, the age of canals lasted only a few decades. Then the invention of the steam engine brought about the railroads that made them obsolete, but not before they had changed the world.

England was a major beneficiary of the canals. Before the canals every pound of coal that came out of a Welsh mine had to be trundled, one wagonload at a time, over rough roads to the factories and homes that burned it. But once some enterprising soul had dug a ditch to fill with water, all these bulk materials—coal and iron ore and grain and everything else—could glide from source to market by the bargeload. On the frictionless canals one tow horse could pull as much as eighty tons of goods, hundreds of times more than it could manage when yoked to a wagon. It no longer cost more to deliver goods than it did to produce them, and industry and trade flourished.

In America, one canal did more than that: It opened most of a continent up to commerce.

When Henry Hudson discovered the river that bears his name, he sailed his *Half Moon* up it as far as where Albany is now, but no farther. Hudson was looking for a Northwest Passage. There wasn't one, and farther upstream the Hudson was no longer navigable.

What Hudson didn't know was that a few score miles away was something more wonderful: the American Great Lakes. But once the United States began to spread westward, a couple of centuries later, the pioneers knew it. The governor of New York State, DeWitt Clinton, had the idea of connecting the river to the lakes, and in 1817 the digging of the Erie Canal began.

The skeptics laughed at what they called Clinton's Ditch. But nine years later it was complete, a channel fifteen feet deep from Albany to Lake Erie, and the whole interior of the continent was open to travel and commerce. The cost of shipping a ton of grain back to the markets along the Atlantic coast fell from a hundred dollars to ten, and the country prospered.

There isn't much left of the world's network of canals. Most of them survive only in fossilized form. You can find a "Canal Street" in most American cities to show where the canal ran before it was made obsolete and filled in. You can notice, too, that many streets coming to the "Canal Street" don't directly meet; that's because when the canal was functioning, they didn't meet at all.

Some stretches of the Erie and a few other American canals remain, kept filled with water mostly for pleasure boating and because they're pretty. In Europe there's more. Tourists can cruise the old canals in France and Germany. In England there are the same sort of cruises, and also at least one less formal one. In summer it is possible to ride a canal boat from London's Regents Park Zoo to Little Venice, a few miles away. The route goes through tunnels under hills and passes long-abandoned docks and warehouses—and once in Little Venice, there's an Underground station nearby to take you wherever else in the city you want to go.

Then there is England's Grand Trunk Canal. It parallels the main railroad line north from London, and you can see it from the train. About all it carries now is pleasure boats and the odd purpose-built houseboat—they are called narrow boats for obvious reasons—for people to cruise and sometimes live in.

The *practical* functions of England's canal network are now few. Like Chicago's old underground freight tunnels, some of the British canals are now used as rights-of-way for communications wires and fiber-optic cables, and that's about it.

But they might do more. Some British engineers argue that the canals could be usefully employed for the very purposes for which they were built, namely for transport of some bulk materials. That sort of transportation would be cheaper and vastly less polluting than the current fleet of trucks that do that job. Of course, it also would be much slower.

However—the engineers point out—that isn't always a disadvantage. Such bulk materials as coal, for instance, don't usually go straight from the truck or train to the furnace. At its destination the coal goes into huge dumps outside the plant, waiting its turn to be shoveled out and burned. But the canal barges would be just as good as a dumping ground on a power-plant lot for holding that reserve. The canals themselves could be the holding dumps.

CANALS ARE QUICK and easy to dig in flat country like The Netherlands. Hilly northern England and Scotland were much more of a problem. If you dig a canal up or down an incline, it won't work; the water will simply run out at the lower end.

To solve that problem, the engineers built the kind of watery elevators called locks. A lock is simply a pair of great watergates that trap a pond between them. Sail into the lock from the downhill side and close the gate behind you. Let water from the uphill side fill the trapped pond. Your vessel rises with the water; when it is high enough you open that gate and sail on, anything up to a hundred feet higher than you were before.

The old English canal system is full of such locks, and there are plenty in America, as well. Chicago has a fine set of locks, and they're easily accessible in summer. Then the Chicago River is full of excursion boats that go out into Lake Michigan for cruises up and down the shore, and every time one of them goes into the lake it must go through the locks.

It is a system of locks that makes possible the Welland Ship Canal, linking the St. Lawrence Seaway with Lake Ontario so that ocean-going ships can bypass Niagara Falls and enter the Great Lakes. If you drive from Buffalo to Toronto you cross it, and it is startling to look out the window of your car in Ontario and see a great ore carrier sailing through the housing developments and apple orchards.

Probably the grandest locks in the world are the ones of the Panama Canal. They are the ones that jack vessels from the Atlantic Ocean up to Lake Gatun, and then let them gently down to the Pacific. They don't do that with a single lock, but with a series of them, and these locks are gigantic enough to accommodate anything but the very biggest of warships and oil tankers. As the ships move through the narrow canals, they are pulled not by horses or mules but by powerful electric locomotives running in tracks along canalside.

I can't say much about the Panama Canal locks from personal observation. It isn't that I've never seen them. Indeed, I spent a year of my life in Gatun, in the Canal Zone. However, that year happened to be the first year of my life—my father took our family to the Zone for a one-year stay six weeks after I was born—and so, although I am sure I was within eyeshot of the locks many times, I don't suppose I paid them much attention.

WHEN I MENTIONED the locks on the Chicago River at Lake Michigan, it may have occurred to you to wonder why a river flowing into a huge lake needs locks.

The reason is simply that the river doesn't flow that way anymore.

Chicago gets its drinking water from Lake Michigan. From almost anywhere along the lakeshore you can see the water intakes sticking up far out into the lake. The water from the intakes is sucked into a modern and efficient treatment plant, and by the time it passes through that it has become drinking water as good as any in the world.

This wasn't always entirely true. Like most big cities, a couple of generations ago Chicago dumped its sewage, untreated, into the handiest large body of water. Which, in their case, was that same lake.

As the city grew, and more and more people were contributing the contents of their toilets to the pool, that got serious. The sewage that went into the lake came right back in their drinking water. What they needed, they decided, was a river to carry the sewage away. They did have a river, all right, but the confounded thing flowed *into* Lake Michigan and thus was no use at all for carrying contaminants away.

So the engineers simply turned the Chicago River around. They dug a canal to connect it with the headwaters of a tributary of the Mississippi, and now it flows backward, away from the lake.

It wasn't a permanent solution. Since cities downstream along the Mississippi weren't pleased about being the recipients of Chicago's sewage, Chicagoans ultimately had to install sewage-treatment plants to clean up their effluent and their river anyway. . . . When they could, that is.

There's one stretch of the Chicago River's South Branch that still resists the best efforts of the cleanup squad. It runs through what used to be the Chicago Stockyards. The slaughterhouses produced an enormous quantity of animal waste, and kept on doing it for years, every bit of which poured into the South Branch.

In his novel *The Jungle* Upton Sinclair described the South Branch in unflattering terms: "Here and there the grease and filth have caked solid, and the creek looks like a bed of lava; chickens walk on it, feeding." A contemporary of his commented that the stench from it "would give a mule hiccoughs."

That was then. It's better now—somewhat. The South Branch is now definitely liquid all the way through, and the smell is almost gone. But much of that burden of fat, blood, and animal waste still lines the bottom of the stream. It is slowly decaying, and as it decays it releases a constant stream of bubbles.

If you are interested in seeing it for yourself, there are occasional sightseeing cruises along that damaged stream; the organization sponsoring the cruises is called the Friends of Bubbly Creek.

What made Bubbly Creek at least cosmetically cleaner—and began cleaning up even the East River in New York (where one of the all-day anglers I used to watch when I lived nearby actually caught a live fish a few years ago)—was the work of a large number of citizens who cared about the environment enough to lobby legislators into passing laws with teeth in them. They're not just in the United States. I saw a bunch of them at work even within the Soviet Union's Kremlin in Moscow, back when the Soviet Union still was the Soviet Union.

YOU HEAR A lot of criticism of greedy capitalists who destroy forests and waterways for the sake of profit; much of the criticism is richly deserved. But if there ever was a regime in the world that ruined whole countrysides, it was the reign of the Soviets. Under their successive five-year plans production was god. Whatever increased production was good; what harm it may have done to the environment—or to the workers!—did not matter.

There are a thousand such cases, but the particular environmental disaster I have in mind here is what happened to the Aral Sea. The sea had existed, large and full of fish and other resources, as far back as anyone could remember, until the middle of the twentieth century. Then the plan for Soviet agriculture decreed an increase in cotton production. Accordingly, vast new plantations were established nearby. Cotton is a thirsty crop, and so these new farmlands were irrigated with water drawn from the rivers that filled the Aral Sea—so much of it that the sea began to dwindle, shrinking to a fraction of its previous size. Old Aral seaports are useless, because they sit on dry land, miles from where the sea now begins. Great stretches of territory that had once been underwater became bare ground, useless for planting and littered with the rusted skeletons of fishing vessels that were abandoned as the water receded.

This was troublesome for the central planners in Moscow. It wasn't that they cared about the sea itself. What concerned them was its economic value. The Aral Sea had been a rich provider of fish; now the fisheries were dying along with everything else around the sea, and so the planners looked for a way of replenishing it. They came up with a daring scheme. There were two great northward-flowing rivers, the Ob and the Yenisey, which were of no particular productive use that the planners could see. So they decreed an enormous public works project. They would build dams and canals to divert the waters of the two rivers to refill the Aral Sea. The whole idea, of course, was a scandal to the world's environmentalists. But it was seldom protested within the Soviet Union itself, where protests generally were rewarded with a free trip to the *gulag* (prison).

That was where matters stood in the mid-1980s. The plans were complete. The rivers would be diverted. Work was almost ready to begin.

It was then, in the summer of 1986, that I happened to visit the Soviet Union.

I had two purposes for the trip. The nuclear power plant at Chernobyl in the Soviet Ukraine had exploded violently some ninety days earlier, and I wanted to begin the research into that terrible disaster which ultimately wound up in my novel on the subject, *Chernobyl*. And the Union of Soviet Writers had invited me to attend, as a foreign observer, the every-five-years convention of their trade union.

The convention was a major state occasion. The meeting was held within the walled-off enclave of the Kremlin itself, in the Great Hall that had once been the courtroom of the tsars, with more than two thousand of the Soviet Union's most famous writers in attendance.

I have to say that I hadn't been particularly eager to attend. I had been in the Soviet Union several times before, largely to lecture on

American writing on behalf of the cultural exchange program of the U.S. State Department. I knew what these meetings were like. There would be a lot of self-congratulation by the officials of the Union, vows of support for the current plan and the directives of the most recent party congress, and endless hours of boring, content-free speeches. Then there would be a lavish banquet, with the kind of food you never saw in Soviet shops, washed down with endless amounts of vodka and Georgian wine, and then we would all go home with nothing changed in any way.

I turned out to be wrong about all that. What had changed things was that Mikhail Gorbachev had come along.

After the grimy, repressive years of Brezhnev and his predecessors, this new man was preaching glasnost—openness—and perestroika—rebuilding, and now Soviet citizens were encouraged to speak up about what was right and try to change what was wrong. And, one after another, here in the Kremlin itself Soviet writers were enthusiastically doing it. They accused the officers of the writers' union itself of keeping themselves in power by means of fraudulent elections. They demanded the right to publish their work freely, without state censorship. And they proclaimed the folly and wickedness of destroying the Ob and Yenisey rivers, with all the plan's consequences in damaging the environment.

This was a very new thing in Soviet society. My translators sat beside me, openmouthed, as astonished as I was. What Soviet citizens had whispered behind their hands for decades was now being said out loud, in an official meeting, with television cameras recording every moment—and, for several long hours, with Gorbachev himself sitting quietly in the room and listening to everything that was said.

He didn't just listen; he *heard*. Before long, there were more and more public protests against the plan for the rivers. It was quietly dropped. And that meeting of the Union of Soviet Writers was where the public protests all began.

Well, those were the Gorbachev years. Everything was loosening up in the Soviet Union. Things seemed to be getting better in every way. But Gorbachev didn't last, and neither did the Soviet Union. Those promising days were followed by violent economic dislocation, pervasive organized crime, and a governmental structure somewhere between empire and anarchy. But at least the Ob and the Yenisey, still intact, continue to flow unhampered to the Arctic Ocean.

* * *

RIVERS ARE IMPORTANT to the human race. It was rivers that enabled the first human civilizations to develop. The place where cities began cwas in Mesopotamia—the word means "the land between the rivers," those two rivers being the Tigris and the Euphrates. The great Egyptian civilization of the pharaohs could not have been without the Nile, and almost equally important were the Rhine, the Volga, the Danube, the Yangtze, the Congo, the holy Ganges of India.

Rivers come in all shapes and sizes, and the United States possesses the full spectrum. There's the immense Missouri-Mississippi system, which we've already discussed. It is not quite the biggest freshwater river on Earth. (That honor belongs to the Amazon in South America, which collects and delivers so much water to the sea that miles from shore, out of sight of land, the water of the Atlantic is still fresh.) But it is still huge and an indispensable roadway for commerce and source of drinking water for all the cities along its stretch. There's the Colorado, with its vast chain of canyons cut by the rushing water over the ages; if you take the 225-mile rafting trip down the Colorado from Navajo Bridge through the Grand Canyon, you're actually traveling in time to the early days of our planet. From the present surface you go down through limestone that was laid down more than 200 million years ago, finally to the so-called Vishnu stratum of 1.7 billion-year-old black schist at the end.

Almost as beautiful, though nowhere near as old, is what is called the Grand Canyon of the Pacific, at Waimea on the Hawaiian island of Kauai. Waimea Canyon is ten miles long, a mile wide, more than a half mile deep. Since Kauai itself is only a few million, not billion, years old, you may wonder how the tiny Waimea River can have done so much in so short a time. Perhaps the answer is simply that it had so much more water to work with; the central mountain of the island, Mount Waialeale, is one of the wettest places on Earth, receiving more than four hundred inches of rain a year.

Even the Hudson dug its way through rock, as you can see when you look from any Manhattan skyscraper at the Palisades on the New Jersey side. Indeed, the Hudson cut more than you can see. During the ice age, when the oceans were lower than they are now, the Hudson cut a channel clear out to what was then the coast. That drowned riverbed is still there on the sea bottom, now a vast underwater gorge, extending ninety miles out from the present shore.

Then there are the rivers that do not dig great valleys for themselves, like the Platte—a mile wide and an inch deep, as they say. And then there is the Everglades.

We think of the Everglades as a sort of tropical jungle, but what makes it so is that it is essentially a river—a very broad and very shallow one. It runs from Lake Okeechobee to south of Miami along that flattest of all states, Florida. It is as much as sixty miles wide, and much of it is less than knee deep.

Or was.

Farmers and the Army Corps of Engineers have done their best to destroy the Everglades, and they've come pretty close to succeeding. The Corps of Engineers has gouged out new channels to straighten the meandering flow of the river, sometimes to make new farmland, sometimes, it seems, just to be neat. Three-quarters of the waters have been stolen, some to irrigate sugar plantations, some to fill the pipes of the mushrooming cities where good northerners go to die.

The result is that there isn't enough flow left to keep the southernmost part of the Everglades wet.

Now that this has become too clear to ignore, the engineers are at last trying to undo what they did a generation ago. Step 1 is taking forty thousand acres out of agriculture and returning them to their natural condition as a swamp; the restored wetland will not only help restore habitat for Everglades creatures but, by sopping up excess nutriment from farm runoff, go on to make the water that many Florida cities drink pure again. If the engineers go on to fill in the channels they dug and thus allow the river to spread out and meander again, much of the Everglades may be restored—but it will never again be what it was before human beings got their hands on it.

IT IS THE nature of rivers to run downhill; that is what makes them flow. If the gradient is gentle, they flow slowly. If it is steep, the flow is brisk, sometimes even violent, as in the whitewater rivers of the western United States. If it is *very* steep, what you get is a waterfall.

There aren't many spectacles on our planet more beautiful than a waterfall, and we have been provided with plenty of them. When they are small they are called rapids, and most rivers have a few of those somewhere along their courses. When the drop is large, they are magnificent.

I have a personal fondness for the waterfalls of Hawaii's Big Island. If you drive north from Hilo you will pass a couple of them, but some of the best can be seen only from the sea. As I said earlier, Hawaii's islands are temporary things. The sea begins carving away at them almost as soon as they are born. Sometimes the erosion slices away great chunks of mountain to fall into the sea, leaving waterfront cliffs. The rivers there, which once flowed sedately into the ocean, now emerge at the top of the cliffs as waterfalls.

One of the most spectacular I have ever seen is in Brazil, just where that country, Argentina, and Paraguay come together in the region called Iguaçú. The river that makes these cataracts of Iguaçú begins on the high plains. Where the tableland ends, the cataracts fall into a scoured-out valley: thus the many cataracts. How many are they? That depends on how much rain has recently fallen on the headlands of the river, so it is worth inquiring before you go what the weather has been. When I visited Iguaçú it was January and the rains had been gratifyingly copious. Looking out over the valley from my hotel room I could see dozens of waterfalls, great and small, popping out of the cliff on the other side.

Some rewarding falls have taken me by surprise, like the Swiss Alpine waterfalls seen from the train between Zurich and the Italian border. Most memorable of them were twin cataracts, from separate sides of a high valley; they actually crossed, or seemed to cross, as they fell. No one in my compartment spoke any of the languages I can manage, and I have never been able to find out if that wonderful display had a name or was only some sort of illusion.

What must be the most famous waterfall of all is that traditional destination for honeymooners, Niagara Falls. Since I was eight years old I have visited Niagara many times, in summer and in winter, and I have always hated to leave. Perhaps the falls are most beautiful in cold weather, when they are cloaked in vast sheets of gleaming ice. In summer strolling along the observation walks is a lot more comfortable, though. Take your pick.

Like Iguaçú, the Niagara River gets its drop from the difference between the elevated plain, in this case the Midwest, and the lower lands here those falling toward the Atlantic coast. Unlike Iguaçú, Niagara's display is scarcely affected by recent weather. It is fed by the outflow of the world's greatest expanse of fresh water, the Great Lakes, and the water just keeps coming.

That enormous steady flow always has been a great temptation to the electric utilities. In fact, for many years a portion of it has been diverted to the little power plant hidden at the base of the falls. What is undiverted remains one of the great sights of the world. It is spectacular from either the Canadian or U.S. sides (though the Canadian Falls are definitely the more beautiful). It is even more spectacular, or at least equally so in a different way, if you board one of the *Maid of the Mist* boats that take you into the river below the falls, so that you can look up at them. You will get very wet in the process, though not as wet as you will get if you venture to that other tourist attraction, the walking tour that brings you on foot almost to the base of the falls themselves.

Like all good things, Niagara Falls is a temporary geological phenomenon. All that rushing water eats away at the rock it falls over. Thousands of years in the past the falls were just as spectacular, but located many miles downstream of their present position. A couple of decades ago a vast rockfall showed how the process worked, as part of the flow suddenly lurched some yards upstream; its debris is still visible at the foot of the falls.

Having the falls migrate upstream does not please the entrepreneurs in the hotels, restaurants, and gift shops, so they asked engineers to put a stop to it. They've done their best to stabilize the falls where they are . . . but sooner or later the water will win.

SO FAR THE water I've talked about has been liquid, whether oceans, lakes, rivers, canals, reservoirs, or waterfalls. But water does come in other states. A good deal of the Earth's water is in the solid state. It's called ice.

The most peculiar thing about ice is that it, as the solid form of water, is lighter than water's liquid state. Ice floats on water. There aren't many substances like that, and that odd trait is one of the things that makes our life possible. Ice is also a pretty poor conductor of heat, which means that when a body of water freezes over, the liquid water under the ice is somewhat protected from the cold air above. Fish can live under the ice. Plants may go dormant, but they don't die. How fortunate we are that this oddball among liquids is also the commonest liquid on our planet.

For most of us who live in temperate climes, ice is what we get out of our freezer and curse at on slick winter roadways. But where the climate is colder—because the area is closer to the poles, or on a high mountaintop—ice accumulates. It falls from the sky as snowflakes, often feathery light, but as layer builds up on layer, their own weight squeezes out their six-sided structure and they become amorphous ice. When a lot of ice accumulates in one place, we call it a glacier. When glaciers merge, and expand, and begin to cover a significant fraction of Earth's surface, we call it an ice age.

Ice ages begin with a series of colder summers. In northern climes the snow cover melts away slowly in a normal year—I remember seeing drifts in Moscow, hidden behind a wall or a building, as late as mid-May. If a slight drop in average temperature allows the snow to last a little longer, there's a feedback effect: The white snow reflects more of the warming sunlight away, the ground stays colder, the snow cover takes longer to melt . . . and then, one summer, it does not melt at all.

That's how our most recent ice age began, in Europe and Asia as well as in North America. In the Western Hemisphere it started on a plateau in northeastern Canada and crept down to bury much of the upper Midwest under a mile or more of solid ice. Along the eastern seaboard it reached what is now New York City's midtown. As the weight of ice grew and moved south, it pushed before it boulders, soil, sand, and everything else that was movable. In places it gouged out long north–south valleys, some of which later filled with water and became New York State's Finger Lakes. In New York City's Central Park, where the bare "basement" rock protrudes above the soil in places, you can still see scratch lines in the stone that were made in the same way, though on a smaller scale, and with the same north–south orientation. The glaciers carried smaller objects, including good-size boulders, along with them. When they melted they dropped the objects where they were; some of those boulders, called erratics, can also be seen in Central Park. Other boulders were left in place, but as the glaciers scoured over them they were spun like tops, drilling out the peculiar, cylindrical potholes that are still visible in parts of the Catskill Mountains.

How much ice was it that did all this? Countless millions of tons. Enough so that the ocean levels all around the Earth fell by many feet, since so much water was now frozen on the continents. Enough so that its weight caused the land under it to sink, and even now in many places is still only slowly springing back to its normal levels.

As I mentioned before, that most recent ice age was not the first the world experienced. Likely it will not be the last, either. For reasons poorly understood (perhaps having something to do with eccentricities in the Earth's orbit around the Sun), from time to time the global temperature falls a few degrees, the summers become colder, and glaciers begin to form.

It has been eleven thousand years since the ice last began its retreat, and the record of the rocks shows that previous warm spells have not lasted much longer than that. Still, at the moment Earth's temperature is going in the other direction with global warming. This trend apparently is related to the human race's newfound ability to generate heat-retaining gases that form a sort of greenhouse in the air. So it is a brand-new phenomenon, and there is no good way to predict what that means regarding a new ice age.

IF YOU WANT to see glaciers today you can go to Greenland or the Antarctic (but I never have, because they are too cold). Or you can visit the Swiss Alps or Alaska.

In my experience, the most comfortable way to visit the Alaskan glaciers is to take a summertime cruise north from Seattle or Vancouver along the inland waterway. You'll see spectacular scenery as well as the occasional whale and, perhaps, a bear or two prospecting for food along the shore. And at the head of some of the lovely fjords you'll see the glaciers.

You can do more than gaze at them. If your ship stops at Skagway or some other port, a helicopter will take you to the top of one of them. Then you can walk around on the ice, where you'll see the trickle of meltwater making little streams and puddles on the surface—that's how Lake Missoula must have begun!—and you'll hear the faint, constant murmur of running water from the larger streams below. Then, back on your ship and at the head of one of the fjords, you may see the consequences of that melting as your glacier calves a little iceberg now and then, with a sound like a cannon shot.

It is normal for a glacier to melt a little at the top where the sun warms the ice and to melt at the bottom, as well, from the sheer weight of the ice. It is also normal for a glacier to crumble at the edge to make icebergs. What is not normal is that at this particular time, perhaps because of global warming, the glaciers are retreating all over the world. Half the ice in Switzerland's alpine glaciers has melted and run off in alpine streams—so much so that corpses, trapped in a sudden freeze or fallen into a crevasse hundreds or thousands of years ago, are turning up. In Africa, the ice cap on Mount Kenya has lost 40 percent of its mass since 1983. In South America, the mountaintop glaciers of the Andes are dwindling so fast that they may be gone in another few decades—with dire consequences for the people who live nearby and depend on their meltwater for drinking and farming.

THE GREATEST CONCENTRATION of ice remaining in the world is on the continent of Antarctica. How much ice are we talking about? Enough so that if it all melted at once, the world's oceans would rise by three hundred feet or so, drowning New York and London and Shanghai and Rio de Janeiro and a thousand other of the world's cities, great and small.

That isn't likely to happen, at least for a good long time. Some melting is definitely going on, perhaps because of global warming. Sections of the Antarctic ice shelves as big as an American state have broken off and floated away. But the main mass of ice does not appear to be in immediate danger, partly because that much would take a very long time to melt, partly because the Antarctic ice cap actually may grow because of global warming. If the oceans warm up, they will evaporate more water vapor into the air. When that water vapor drops out again as precipitation over Antarctica, it will fall as snow and thus actually *add* to Antarctic ice.

I have to tell you that nothing I say about Antarctica comes from personal experience. I've never been there. I don't plan ever to go, either.

This isn't because I wouldn't like to see it for myself. I really would—just as, for instance, I would really like to make a good, old-fashioned bonfire of the leaves that accumulate on my lawn every autumn. In both those cases I don't do it because my environmental conscience says no. Antarctica is too fragile an ecosystem to tolerate even as much tourism as already visits it each year.

There is also the not-inconsiderable fact that Antarctica is not the safest place in the world to visit. People do die there, sometimes by falling into crevasses in the ice. That doesn't usually happen to tourists, because they are seldom permitted to wander into dangerous areas. But when it does happen, regardless of whom it happens to, the working scientists have to take time off from their researches to mount rescue expeditions. I do not wish to discommode any of them in that fashion.

Well, let's be honest. There's another reason why I don't intend to go to Antarctica, or anywhere near it. It's *cold* there.

I'VE TALKED ABOUT all the waters you can see, whether in liquid form or ice, but there's a very important lot that you don't see.

For example, there are the world's numerous underground aquifers. An aquifer is sometimes described as a buried lake, but that gives a wrong impression. An aquifer isn't an underground pool of water. What it is is a bed of some permeable material, such as fractured rock, gravel, or coarse sand, which has a layer of something as impermeable as sandstone or clay below it and another such layer on top. If there are cracks in the layer above it, as there usually are, rainwater seeps in. If there is enough seepage, ultimately it fills the gravel bed with water caught between the grains. Aquifer water may be thousands of years old—there are, for instance, good-size aquifers deep under the Sahara, where it has hardly rained at all for many centuries—which is why it is sometimes called fossil water.

Then, when people dig down through the top layer, their digs become wells and they can pump water out of the aquifer. In many parts of the Southwest's cattle country, you will see windmills rising here and there from the plain. That's what they are doing, pumping water into troughs for the grazing livestock. (At present you also will notice that many of those windmills aren't turning. That is because, with gasoline so cheap, it is less trouble to run the pumps with a small gas motor. As the price of gas rises, this may change.)

Sometimes no pumping is necessary. Those sandwiched underground layers of watery gravel start out horizontal, like all deposits on Earth, but—like all strata—they are at the mercy of tectonic forces. If one end of the aquifer is elevated above the other, then the contained water would like to run downhill. If it is securely trapped by the impermeable layers above and below it, it doesn't have anywhere to run to—until some fissure at the downhill side gives it an escape route and it becomes a spring. Or until some well-digger drills a hole through the crust. Then the pressure of the higher-up volume of water turns that well into a fountain, sometimes spurting ten feet or more into the air. It is then called an artesian well. In some cases the artesian flow is great enough so that a turbine can be hooked to it and used to generate electricity.

But there aren't as many of those as there once were, because a lot of the world's aquifers are close to being pumped out. Even the great Ogallala aquifer, which underlies half a dozen western states and has provided irrigation for hundreds of square miles of agriculture, has been widely, and in some cases perhaps terminally, exploited. As you fly over these states you will see perfect little circles of green here and there on the dry land. They are made by Ogallala water, pumped up from the aquifer and spread over the crops by a revolving arm of sprinklers from the central well. Or, from the ground, you may see skeletal rows of sprinklers on wheels, a dozen feet high and a hundred yards long, doing the same thing.

But some of this irrigation has a limited future, because the great Ogallala is being pumped faster than it is replenished by rainfall. In parts of New Mexico and elsewhere, the water table has dropped ten feet or more in the lifetimes of living farmers. Some of those pumps are now sucking mud.

In other places, particularly in shore areas like Mexico's Baja California Peninsula, aquifers have not gone dry but have become nearly useless. So much freshwater has been removed that sea water is seeping in, and for many crops the water has become too salty to be used.

THERE IS ONE more vast mass of water that we can see only in part, although it is all around us. That is the water vapor in the Earth's atmosphere, millions of tons of it. The reason we can't see most of it is that water *vapor* is as transparent as the air itself. It is only when the vapor condenses into drops of liquid water that we can see it at all. Then it appears as clouds or fog (which itself is only a cloud that has come down to ground level).

You can see for yourself the difference between water vapor and cloud next time you make a pot of tea. When the teapot is boiling furiously, what comes out of the spout is invisible water vapor. It is only after it has had a chance to cool into liquid droplets, an inch or so from the spout, that it becomes visible steam. Clouds are a sort of low-temperature steam. They occur when warm air, saturated with water vapor, cools either by being elevated as it goes up a slope or by contact with a colder air mass, as in a weather front. The vapor condenses into droplets; the clouds appear. That is how the flat-bottomed cumulus clouds you see in a summer sky are formed. There is an upward current of warm air; when it reaches a sufficient height, the cloud forms, the flat bottom showing the precise altitude where the cooling takes place.

It is the water droplets in a cloud that make one of the prettiest of Nature's sights, the multicolored rainbow.

Why does the rainbow have colors? Because white light, or even the off-white light that comes from the Sun, is not a pure color. It is a mixture of all the colors there are. The water droplets in the cloud, like the prism or diffraction grating of an astronomer's telescope, break this white light into its colored components. Conventionally, those components are, in order, the colors red, orange, yellow, green, blue, indigo, and violet. (If you have trouble remembering them all, think of the mnemonic, which is the name of a mythical person called Roy G. Biv.)

Here, too, as in gravitation, it was Isaac Newton who turned optics into a science. He proved that light is a mixture of other colors by splitting it into its components with one prism, then reconnecting them with another—once again producing the white light he had started with. He also, however, did something else that I wish he hadn't. He included indigo as one of the component colors.

I've never been able to see indigo in a rainbow. I don't think anyone else has, either, really. The thing is that Isaac Newton was a religious man. Other experimenters thought there were only four colors in the rainbow. Even Newton's own assistant could, like me, see only six. Newton was obdurate. It was for religious reasons that he thought that there should be exactly seven colors—the mystical number—so he squeezed the mythical indigo in between the blue and the violet. And, since he was *Isaac Newton*, his opinion became gospel.

Unless you are in an airplane at high altitudes, and only if the conditions are just right even then, you never see the whole of a rainbow. From the surface you can't see a rainbow at all unless the Sun is behind you and not more than 42 degrees above the horizon, either rising or setting—and unless there happens to be just the right kind of a cloud in just the right part of the sky, near the horizon. Then, if you're lucky, you may see that magnificent arc of color that we call a rainbow.

But you don't see the whole thing, because the rainbow isn't just an arc—or a "bow." It is a complete circle. Why we don't usually see the entire circle is because the center of that circle is at what is called the antisolar point—that is, the point in the sky where the Sun is going to be twelve hours later or where it was twelve hours before. Naturally you can't see that point from the surface. You have to get really high up for that purpose, because the antisolar point is on the other side of Earth, and the planet itself is in the way.

So it really isn't worthwhile to go looking for the leprechauns' pot of gold at the rainbow's end. Like every other circle that has ever existed, a rainbow doesn't end.

Chapter 7
WONDERS UNDERGROUND

Caves and Tunnels

In all my talk about the crust of the Earth there's one aspect I haven't mentioned, and that is the holes that are in it. Of these there are plenty, both natural and man-made, and some of them are spectacularly beautiful.

When the holes in the planet are natural we call them caves, and they have been notable tourist attractions since the beginning of recorded history. Some of the most interesting come from myth and fable—the caves under King Minos' palace on Crete, where Theseus pursued the Minotaur, for example; moreover, Odysseus' Cyclops was a cave dweller, as were the sirens who tried to entice his crew called from a grotto on the Italian shore. There's not much chance of visiting these, however. If they ever existed, no one can say exactly where.

There is one fabled cave, however, that really does exist. Moreover, it can be visited by anyone who happens to be in Athens, Greece, with an extra day on his hands. That's the one once occupied by the Delphic Oracle, on the flank of Mount Parnassus. The Delphic Oracle was perhaps ancient Greece's top-rated psychic, and all right-thinking Greek kings and generals climbed the mountain to get her opinion on what to do next. Getting there was a chore, in those days before paved roads and tour buses, but what was even harder was figuring out the Oracle's advice once given, since it was always cloaked in allusion and parable.

These days getting to the general neighborhood of the cave is a snap. The road from Athens to Parnassus passes through some beautiful and historic territory, including the Plain of Marathon. (Remember Marathon? Big battle? Runner brought news, starting a fad?) Once you've driven up the mountain, though, there are still a daunting number of steps to climb to reach the site itself. However, there's a museum at the level where the steps begin, containing some fine statuary and what is said to be the actual omphalos, the stone that the Greeks regarded as the belly button of the world. So it may not be worth making that last climb, especially since the Oracle doesn't seem to live there anymore.

Which may be just as well. Some current archeologists, poking around the site, discovered ancient fissures in the rock—which, they believe, emitted noxious fumes; the Oracle is said to have sniffed them before making her appearance to supplicants, and so when she proclaimed her cryptic revelations she may have been simply stoned out of her mind.

Then there are the caves visited by human beings in prehistoric times. Notable is the cave of Lascaux, in the French Dordogne, where the artwork of primitive human beings is preserved today. Aurochs, deer, oxen, and horses are there at Lascaux, incised on the walls and painted in shades of reds, browns, and black. However, you can't see them for yourself anymore. Those cave paintings survived unharmed for fifteen thousand years—until their accidental discovery in 1940—but then things went downhill. The warmth and lights brought by the hordes of tourists that followed—and especially the moisture in their collective breath—dimmed the colors and brought about the growth of an unpleasant green fungus. The cave was closed to the public in 1963.

Happily, however, the locals wanted to keep those tourist dollars coming. They have recently completed a copy of the cave—made with more durable paints and controlled air circulation—open to visitors nearby.

Cave drawing appears to have been a major preoccupation in Stone Age times. There are at least a half dozen other such sites in Europe, including the one called Cosquer, on the Mediterranean coast. Cosquer has some of the earliest cave paintings known, but it's not at all easy to get to. It was discovered, in 1991, by the professional scuba driver, Henri Cosquer. The land has settled since the time those pictures were drawn, and to reach the cave Cosquer had to traverse an underwater passage some five-hundred feet long.

NATURAL CAVES OCCUR in many places, caused by many different processes. Some arise from volcanic action (often from buried lava tubes like those of Kilauea), others from volumes of gas trapped in rock as it solidifies (remember Arecibo's giant bubbles). Most of the most impressive ones, however, are made by water. Some of those are grottoes carved out by waves and river currents; most are simply holes dissolved away by the slow seepage of groundwater through minerals like common salt or, especially, limestone. Carbon dioxide from the atmosphere dissolves in rainwater as it seeps into the ground, forming carbonic acid. As acids go, it's a weak one, and usually very dilute in groundwater, as well; but given enough time, it can dissolve away vast quantities of rock.

As vast, for example, as Kentucky's Mammoth Cave, which is one of my personal favorite caverns. It's in Cave City, ninety miles south of Louisville; if you happen to be going to the next Kentucky Derby, stay over a day for the trip to Mammoth Cave. Be prepared to do some walking. There are more than three hundred miles of connected caverns in and around Mammoth. Although you won't be expected to cover all of them—you couldn't if you wanted to, because much of that range is inaccessible to amateurs—there's still much to see. There's an elevator to take you down (it's over 350 feet to the very bottom, though you don't go quite that far), and a coffee shop and souvenir shop where the elevator lets you out; then you follow a guide to explore.

In Mammoth Cave the temperature remains at a steady 54 degrees Fahrenheit, winter and summer, so you may want a sweater, as well. You won't be the first visitor. That honor belonged to some Paleo-Indians five centuries or more before Christ; they mined gypsum from the cave, and the mummified body of one unfortunate miner, trapped in a rockfall, was found when more recent explorers began to check it out. During the War of 1812 American troops used it to store gunpowder, and to make it homey they installed plumbing, of a sort—hollowed-out trunks of the tulip tree, which are still there. Now the cave has only recreational uses, but they are first-rate.

Another cave I am fond of is much smaller, but hallowed by association with my favorite American novelist. That's Tom Sawyer's Cave, just south of Hannibal, Missouri.

In his story, Mark Twain was pretty faithful to the physical circumstances of the cave. The characters, of course, were imaginary. Such things as the spot where Injun Joe died, trying to carve his way out of the cave with a pocket knife, are not to be taken as objectively real, but a guide will be happy to show you what *could* have been that spot, and the fact that it's imaginary doesn't make it much less interesting.

If what you want in a cave is sheer beauty, the loveliest one I know is the wave-carved cave on the island of Capri, Italy, that is the famous Blue Grotto. Getting there is half the fun. You take a motor launch from the port to a point just outside the grotto; there you switch to small boats, because the grotto's entrance is slowly sinking and there's little headroom as you go in. (The guides are good about keeping tourists from falling in, but you'll need to be fairly spry to make the switch.) Once inside you are floating in breathtaking splendor. The only illumination comes from the outside sunlight filtered through the water itself, producing a luminous azure glow more beautiful than you can imagine. Even the surly old Roman emperor Tiberius loved the grotto, and you will, too . . . but don't wait too long for your visit. The island of Capri is slowly subsiding. Even now the entrance to the Blue Grotto by boat is impossible at some seasons, as my wife found out one recent January when she tried to get in, and the time is approaching when only swimmers will be able to enter.

NEARER TO HAND are the chains of strikingly beautiful caves that line Route 81 in Virginia. I haven't managed to see them all, but I well remember the Luray Caverns, with its "Wishing Well" to throw your spare coins into. (About a half million dollars has been fished out over the years and donated to charities.) Then there are Shenandoah Caverns (with their spectacular and fancifully named chambers like the Oriental Garden and the Grotto of the Gods); Skyline Caverns (noted for its Fairyland Lake); the Natural Bridge Caverns, deepest in the East, and with the Natural Bridge itself, called by Thomas Jefferson "undoubtedly one of the sublimest curiosities in nature" right next door.

However, what is by all odds the most magnificent of American cave systems is Carlsbad in New Mexico, a goodish drive north of El Paso, Texas.

Carlsbad is the only American national park that is completely underground. There is a lot of it. The Carlsbad complex of caverns isn't the biggest in the world—that honor belongs to Borneo's Sarawak Cave, and even Mammoth Cave extends over a longer distance underground if you count all its connected systems—but Carlsbad descends over a thousand feet, and its labyrinth of caverns covers some thirty thousand acres.

Primitive humans seem to have missed Carlsbad Caverns, but it was not completely uninhabited. Its underground lakes contain a population of cave fish, blind little creatures about four inches long that find their food by taste and touch, and it is especially populated by vast swarms of Mexican free-tailed bats. One of the great sights of Carlsbad is to stand outside it at sunset and watch the creatures swarm out, darkening the sky, to hunt their prey all night and return to roost in the morning. (But don't get too friendly with the bats, some of which may carry rabies.)

THERE ARE MANY other great natural caves in the world, but those are the ones that have given me most pleasure. Still, they have their limitations. Even Carlsbad is only a fifth of a mile deep. Natural caves don't go much deeper than that, because the weight of rock over them squeezes them out; even granite will flow under the conditions of heat and pressure that occur deep inside the Earth's mantle.

For deeper caverns you have to go to the ones human beings themselves have made. The deepest of those are the mines, some, as in certain African gold mines, going two miles down or more.

The trouble with visiting mines is that they are generally uncomfortable and frequently dangerous. Coal mines are particularly so; "coal gas"—methane—seeps out of the beds, and any little spark may set off a lethal explosion. (That is not news to anyone who reads a daily paper; coal mining is one of the most hazardous of human occupations, and several hundred miners die in such blasts each year.)

Fortunately for us spectators, we don't have to take the chances the real miners do. There are a couple of museums—such as the Museum of Science and Industry in Chicago and the Deutsches Museum in Munich, Germany—that have gone to the trouble of re-creating coal mines on their premises so that any of us can see what one looks like in comfort. (Go early, though. They are popular, and there's always a line.)

A nicer kind of mines to visit are the ones that produce salt. They are not only beautiful (the crystalline salt looks like diamonds) and safe (some even have been used to provide natural air conditioning for people suffering from respiratory diseases), but some of the best of them are located right in major cities. There's a big one under Cleveland, Ohio, and one that has become a major tourist attraction in Cracow, Poland.

In Kansas City, Missouri, a different kind of mine underlies the city. It is a large granite reef, quarried for decades to provide the stone that built the city. Because the reef lies almost parallel to the surface of the ground, the removal of all that rock has left a network of subterranean tunnels and chambers—the collection has been nicknamed Subtropolis—that are now used for business offices, manufacturing, and storage.

Mines have been pressed into service for other purposes, as well. One of America's first prisons was a copper mine. In 1773 the Founding Fathers, looking for a place to put their miscreants, decided they were best off underground. They called the place New Gate (after London's infamous old Newgate Prison). As long as the prisoners were going to live and sleep down in the mine, the authorities reasoned, they might as well dig a little copper, too.

In the event, that didn't work out very well. It was hard to keep an eye on all the inmates in the dark, cramped passages of the mine, and a lot of the prisoners did their best digging on escape tunnels.

AS THE SURFACE of the planet gets filled up and more and more expensive to build on, particularly in cities, there's more and more interest in going underground.

The name for this practice of building things below the surface of Earth is terratecture, and it has been used for everything from auditoriums and shopping malls to private homes. Since the temperature just below the surface of the earth changes little with the seasons, terratecture is most popular where the climate is hostile to shirtsleeve enjoyment—in Minneapolis, for example, where a structure on the campus of the University of Minnesota is largely underground. If you watched the television broadcasts of the hockey games from the 1994 Winter Olympics in Lillehammer, Norway, the five-thousand-seat amphitheater they were held in was a man-made cavern buried a hundred feet underground. And there are fair-size shopping malls under the offices and hotels in the heart of many cities. Often they're in conjunction with railroad or subway terminals, as in Munich, Germany, and Tokyo, Japan.

In times past tunneling for concealment was particularly popular with conspirators or oppressed people—where did you think the term "underground organization" came from?—as witness the Christian catacombs in Rome and Kiev, where worshippers hid out from pagans and entombed their dead (whose skeletal remains still show in some of the niches).

Another good reason for going underground is safety. The British battles of World War II were largely commanded from underground warrens, owing to the annoying German habit of bombing everything on the surface. In Uxbridge, just west of London, 11 Group of Fighter Command had its main war room sixty feet underground. Aircraft spotters and radar stations telephoned in the locations of Luftwaffe aircraft coming toward the city; in the underground chamber the commanders of the group ordered up Spitfires and Hurricanes to defend it. Now a tourist attraction that is open to the public, it has been preserved just as it was on the hottest day of the Battle of Britain, September 15, 1940. While in London itself the subterranean Cabinet War Rooms, from which Winston Churchill oversaw the entire course of the war from the fall of France to V-E Day, has also been preserved and opened to visitors. In Moscow Joseph Stalin and his Politburo also took shelter underground, inside the Kremlin station of the then-shiny new Moscow subway system. They even held their annual party congress there in 1942, when German tanks had the Kremlin towers in the sights of their cannon, but if there is any marker to commemorate the spot I have not found it. And of course in all warring cities that possessed subway systems, their stations became instant bomb shelters for the civilian population.

Even Paris goes underground on occasion. I. M. Pei's giant aboveground glass pyramid before the Louvre is balanced by a similarly shaped underground vault that serves as entrance hall to the museum, and the newest highway bordering the Seine is all underground. And—remember the novels with chases in the sewers of Paris? There's a tour that takes you there, leaving from a spot on the Place de la Concorde.

To get a good look at the complexity that lies under a major city, visit Consolidated Edison's little museum on the corner of East 14th Street and Third Avenue in New York. It's a mole's-eye view of just what is really there at that intersection, though hidden from sight. And if you then walk a block east and then in either direction up and down Second Avenue, reflect that you are walking over one the world's most expensive holes. Several miles of tunnel are under your feet, once intended to be a great Second Avenue subway line, then abandoned as costs escalated. (Not unlike that other great American abandoned tunnel, the partly complete Superconducting Super Collider in Waxahatchie, Texas.) Most cities have abandoned warrens of this sort hidden away under the sidewalks—there's another at the intersection of Flatbush and Atlantic avenues in Brooklyn, souvenir of a failed attempt at a pneumatic underground rail system.

One of London's forgotten underground digs found a way to make itself remembered a few decades back. It, too, was intended as a "pneumatic dispatch tunnel" when entrepreneurs excavated it around 1870. It was a working shaft, not simply an experiment. For a time it actually carried mail all the way from Marylebone to the main post office, near St. Paul's Cathedral. Technically it was a success. But it was an economic failure, and so it was abandoned and forgotten—until 1928. Methane gas had seeped into the tunnel. Some chance spark ignited the gas, and the whole thing blew up, startling the unsuspecting city. (Fortunately there was little damage, except to the pavings along its route.)

FINALLY, THE SUBWAYS.

Almost every major city has them, now that even Los Angeles has opened its own. They come in all varieties. The oldest subway line in the world, still running, is in Budapest, Hungary. When it was built the turns in the tunnel were sharp and frequent. They still are, and so in that ancient tube the individual cars are only about half as long as in the newer stretches of the system. London's Underground is the deepest (and the one that requires the longest hikes to change trains). Tokyo's seems to be the most crowded; at rush hours special "pusher" guards are on hand to shove passengers into cars that, in most systems, would already be considered full. Washington, San Francisco, and Singapore have some of the most modern.

New York's system came along after Budapest's, but its first routes are now about a century old. A plaque at the entrance to the City Hall station on the Lexington Avenue line commemorates its opening. In its first incarnation, from there the subway went north to Grand Central, cut westward along 42nd Street to Times Square, then north again along Broadway.

You can still see traces of that original line. The trains from City Hall now go much farther north than Grand Central and on the Broadway line much farther south than Times Square, and they aren't directly connected anymore. The link along 42nd Street is now a separate shuttle. But as you get off the shuttle at Times Square, parts of the curving tracks that connected it to the Broadway line are still visible. They're visible at the other end, too, if you know where to look. As you take the Lexington Avenue downtown local from Grand Central, off to the right you can get a brief glimpse of the curving rails, now abandoned, that once carried the trains over to join what is now the shuttle line.

Chapter 8
TIMES PAST

Fossils and Archeology

We've gone north and south, east and west, up and down. What's left for us to explore is that fourth dimension of Einstein's continuum, namely time itself.

For that we would like to have a time machine, of course. As a matter of fact, we do have one—the universal one that remorselessly carries each one of us twenty-four hours into the future every day. There's no reverse gear on that, though. It won't take us to the past.

However, all is not lost. Through the workings of blind chance a number of spectacles from the past have survived for us to view today, the most important of which are called fossils. Some of the places where these are to be found are hard to get to in the spatial dimensions, but not all. At least one of the very best is preserved for our viewing right in the heart of a great American metropolis. It's a smorgasbord of fossils, preserved for your entertainment, and it is called the Rancho La Brea Tar Pits.

MOST FOSSILS ARE underground and have to be dug out to be enjoyed; La Brea's are right there on the surface. And, if you're in Los Angeles, La Brea is as easy to get to as city hall; the Wilshire Boulevard bus will take you almost to the door. Not only that, but it's a great spot to visit if you happen to be traveling with someone who doesn't share your interest in science. Los Angeles's art museum is right next door, and the shopping paradise of Wilshire Boulevard's Miracle Mile not much farther. When you've done all the viewing you care to, the famous Farmers Market, no more than a block or two away, can offer you any kind of lunch you desire.

La Brea's fossils aren't very old. There are some uncertain indications of a few specimens from a million years ago, but the identifiable ones seldom go back more than 36,000 years. In those days the site was a bog of oozing asphalt and pools of stagnant water. Thirsty animals came by for a drink out of the pools and found themselves caught in the sticky tar. They died there, and the tar preserved them.

When you count the different species represented at La Brea, you will find that there's something wrong. There are about ten times as many carnivores as herbivores; that can't have been the way it was in life, because what would all those carnivores eat? The answer is probably that whenever some sloth or mammoth or whatever got itself stuck in the tar, it did not go gently into that good night. It probably made a huge racket as it tried to escape, which probably attracted every hungry flesh-eater within earshot. Of those there were a lot: now-extinct species like saber-toothed cats, dire wolves, hyenas, and American lions, as well as representatives of species that are still around, like coyotes, pumas, and bears. The herbivore species included camels, horses, mammoths, mastodons, and ground sloths—all gone now, except for the ones like horses and a few camels that were reimported from Europe in historic times.

All in all, several hundred thousand specimens have been unearthed in La Brea since the days when it was first dug, in the 1900s. The best of them are right there for you to see. The tar pit itself is open for inspection, now housed in a concrete shelter, and just across the park there is a fine little museum to show, among other things, what these great vanished beasts looked like when they were alive.

THERE'S A SIMILAR assortment of extinct American beasts in the state of New Mexico, right between the cities of Clovis and Portales near the eastern border of the state. It's in a place called Blackwater Draw.

These days Blackwater Draw is just a depression in the ground, as bone dry as all the land around it. Twelve thousand years ago, though, it was a spring-fed freshwater lake, and the only one of its kind for miles in any direction—there were other water holes, but they were all either salty or bitter. Therefore, the waters of Blackwater Draw were what the local populations of mammoths, mastodons, ground sloths, and other herbivores—and, as at La Brea, of the carnivores that preyed on them—sought out to slake their thirst.

No human remains have ever been found at Blackwater Draw. None ever have at La Brea, either, but that's hardly surprising; the first humans had yet to immigrate from Siberia when La Brea was most active. Blackwater Draw, though, dates back only about twelve thousand years, and there's no doubt human were somewhere around at that time because the draw is rich in their artifacts.

When you come through the main entrance to the Blackwater Draw site, the first thing you see is a large sign that says:

DO NOT DISTURB THE RATTLESNAKES

That's good advice, but don't let it keep you from walking the trails. If you miss one of the guided tours and walk the site by yourself (as you can, taking with you the little guide pamphlet they will give you at the visitors' center), you should stay on the trails. It isn't just because of the rattlesnakes. Blackwater Draw is an active scientific site, with digging going on all the time; you need to stay out of the way of the working scientists.

Apart from the fossils, Blackwater Draw has another great distinction. It has given archeologists samples of two of the most useful inventions of early man. One is a fluted stone spearhead, generally made of chert or obsidian, which has made the nearby town of Clovis famous by giving its name to these Clovis points. Long and razor sharp, Clovis points are hafted with a long flake chipped out of each side to fit the end of a spear. Point and spear are joined to each other in various ways—at one dig in Oregon the points seem to have been glued to the shaft with melted amber—but they were a great invention, apparently used successfully for killing many kinds of animals, up to and including mastodons themselves.

The other great invention, which came along a little later, was a throwing stick called an atlatl. The advantage of the atlatl was that it gave a hunter the chance to inflict a mortal wound while still standing a bit away from his presumably unfriendly prey. The things worked. To confirm that this was so, scientists have taken copies of atlatls and Clovis points to Africa and tried them out on the nearest thing we have to the mastodon, an African elephant. (They tried them out on *dead* African elephants; they were scientists, not poachers.) And, yes, they found that the points were indeed capable of penetrating the armored pachyderm skin, though lethally at only a few places. Nevertheless, they were indeed used for that purpose, for Clovis points have been found actually within the body cavities of mastodon skeletons . . . though it is possible that the animals survived that first attack and carried the points away, embedded in their bodies, to die later and somewhere else. Which would seem to have been really bad news for the hunters who had put them there.

You aren't allowed to take souvenirs home from Blackwater Draw. But if you want a Clovis point or a later Indian arrowhead to keep as a paperweight, they turn up rather frequently along the Southwest dry lands. I've never found one myself, but my friend Jack Williamson has. He says the best time to look is after a brushfire or a heavy rain, when they may appear on the surface of the ground, and the best place is an abandoned farm, where the ground was once plowed. If what you find is an arrowhead, keep it. If it's some sort of stone tool, including a Clovis point, show it to someone at your local museum; if it's common, it's yours, but if it turns out to be of special scientific interest, you should let the scientists have it to study.

WE'RE GETTING A little off the track. I was talking about fossils, which is to say the things that are left for us to study of what once were living creatures.

There have been a lot of those, in the 3.5 billion years since the first known life appeared on our planet. Some scientists have calculated that if you added up the weight of every living thing that ever existed on Earth, their total combined mass would be just about equal to the mass of the planet itself.

Of course, almost none of that great volume of living things is still around. Most of those creatures were eaten by some other creatures, which were then eaten by still others in turn. Or they simply died and rotted away, releasing their chemical constituents to nourish some new life; Nature invented recycling long before there was a Sierra Club. But an infinitesimal fraction of those creatures did somehow leave traces of themselves that survive to the present day. Those traces are what we call fossils.

Fossils come in all varieties. Amazingly, sometimes the creatures remain nearly intact. Perhaps they were frozen, like the mammoths that have been melted out of the Siberian ice. Perhaps they were trapped in a bog and, covered over in an oxygen-free environment, were preserved for thousands of years. (That period has to be at least ten thousand years for the remains to be considered a fossil; less than that, it's only bones.) Rarely, there are whole-body preserves, though only of quite small creatures, for far longer than that. There is a regular trade in insects from millions of years ago that were caught in the sticky sap of trees, which hardened around them and became amber, ultimately to be cut and polished and made into some lady's necklace; I've seen such bug-containing jewelry for sale in Rome's pricier shops for a few hundred thousand lire. These are by far the oldest whole animals that anyone is likely to find. And sometimes chemical changes have removed all the organic material from an organism—generally vegetation in this case—and replaced it with hard, lasting stone that faithfully reproduces the shapes of the original. Thus Arizona's Petrified Forest.

When we think of fossils, what most of us are thinking of is bones, preferably wired into a recognizable shape and mounted in a museum. However, the most common fossils, comprising 90 percent of all those found, come from animals that don't leave bones, because they never had any bones to leave. These are the invertebrates (which simply means "without backbone"). What they do have to leave for later generations to find are their shells, which are made of lime or chitin—lime being what we find in an oyster shell, chitin the hard, fingernaillike material of a cockroach or a trilobite. Trilobites are a favorite fossil. They look like a cross between a roach and a horseshoe crab, and their best feature is that they are easy to find because there are so many of them. They were around for 300 million years, first to last, and they flourished in large numbers. What's more, they molted periodically, like crabs, and so each trilobite left several distinct shells.

There are even some kinds of fossils that never actually were any part of a living thing—casts, for instance. Sometimes the trapped animal has completely rotted away and disappeared, but whatever covered it—silt, sediment, volcanic ash, or whatever—has retained its shape. If you pour plaster into the space left by the vanished animal, you get a casting of the original, often in astonishing detail. (Interestingly, the plaster mines under Paris's Montmartre section, the place that gave plaster of Paris its name, are themselves rich in fossils.) Perhaps the oddest of recognized fossils is the tube-shape one called Diamonelis, which translates as "the Devil's corkscrew." Diamonelis is a hard, shiny-white mixture of glassy particles and fossilized vegetable fiber, running about six feet long or so. For many years paleontologists wondered just what sort of creature left that enormous shell. When they did figure it out, it turned out that Diamonelis wasn't any sort of creature at all. It was the remains of the linings of animal burrows, excavated by a prehistoric relative of the beaver.

Then there are footprints.

There are something like four hundred finds of prehistoric footprints in the world—a large number at first impression, but really rather small when you consider that there is hardly a square inch of land area that hasn't been trodden over and over by endless animal feet, over the last few hundred million years. Fossil tracks are formed in the first instance when some ancient animal walks across a surface of soft earth or mud. Then those footprints have to be preserved, perhaps by being covered over almost at once by a sudden sandstorm, or a fall of volcanic ash, which is then repeatedly covered over by other falls until it is buried so deeply that it is transformed into rock.

One good set of these fossil footprints is in Dinosaur State Park, at Rocky Hill, Connecticut. Another is in Dinosaur Valley State Park in Glen Rose, Texas, southwest of Fort Worth. I haven't been there myself (yet), but I've seen some of the tracks, which have been excavated and are now on view in the American Museum of Natural History in New York. The one that is called the dinosaur megafreeway is viewable at Dinosaur Ridge National Landmark, at Morrison, Colorado, west of Denver. The landmark is only one section of the "freeway," which, 100 million years ago or so, was the coast of a great inland sea; it is in the beaches and wetlands of this old coast that the dinosaurs left their tracks.

PEOPLE WHO GET interested in looking at fossils sometimes get the habit, and then they may move on to trying to find some for themselves. If that interests you, there's nothing to stop you from trying your luck. There are animal fossils in every state of the union (though not very many in Hawaii, which is simply too young to have any really ancient fossils, or in West Virginia, whose fossils are mostly vegetation preserved in coal seams).

If you do go fossil hunting, remember that you may well need some property owner's permission. This is particularly important in the rather unlikely event that you find something big and/or rare; some fossils have become pretty pricey—in the millions of dollars for, say, a fairly complete Tyrannosaurus rex—and if you do find one, there is not much chance that you'll be allowed to keep it. On the other hand, nobody will care much if, for instance, you kick up an ancient shark's tooth on a beach. If that interests you, a good place to seek one is along the Atlantic coast beaches south of Annapolis, Maryland, where they wash out of the clay banks landward of the beach and almost any warm Sunday brings out a few families to scuffle their feet through the sand.

That sounds like a pretty haphazard way of trying to find ancient relics, but many a priceless fossil has been turned up in pretty much that same casual way. It's how Mary Leakey discovered her first australopithecine (you pronounce the name by starting as if you were going to say "Australia," then finish with OH-pith-uh-sine.) While strolling across an African plain in 1959 she noticed a bone sticking out from the sand, then got out her trowels and picks to exhume the rest of the creature. It's how the eleven-year-old girl named Mary Anning, the world's first systematic fossil hunter, located her thirty-foot ichthyosaurus on the beachs of Lyme Regis, England, in 1910. There are new high-tech ways of finding fossils—sonar, ground-piercing radar, magnetometers, even gamma-ray counters. (Some fossils concentrate minerals that give off radiation.) But a simple eyeball scan of hopeful ground still works.

Where do you find likely ground? A good place is where there has been recent erosion, such as a windblown desert or a streambed. Big floods, like the 1993 Mississippi River one, scour out a lot of surface soil and expose some interesting specimens. Railroad or road cuts are worth a look, particularly abandoned rail rights-of-way, because you don't have to dodge speeding eighteen-wheelers. Mines, building excavations, and quarries may well turn up fossils; my friend Stan Skirvin tells of climbing around coal faces with his kids when they were small. It was in an abandoned open-pit coal mine, he says, that his eight-year-old son came across a distinct tree trunk and finally understood just how coal was formed.

If you want to increase your odds of finding fossils, the best thing is to go where the fossils are. There's Dinosaur National Monument, on the border between Utah and Colorado. That requires a fairly complicated trip—perhaps flying into Vernal, Utah, or driving there on U.S. 40—but many tourists do make the trip, and it is set up for their enjoyment.

In the Midwest, so is Mazon Creek, forty square miles of fossil-bearing ground located an hour or so south of Chicago. You won't find dinosaurs at Mazon Creek, because its fossils are older than that; dinosaurs had not yet come along when its ancient fish, ferns, scorpions, jellyfish, and the like flourished there.

The Hagerman Fossil Beds National Monument in south-central Idaho, along the Snake River near the town of Hagerman, also doesn't have any dinosaurs, but for the opposite reason. The Hagerman fossils are too young for dinosaurs, but they have yielded what is called the Hagerman horse, the direct ancestor of all the present equine varieties, from donkeys and zebras to Thoroughbreds.

If you like scuba diving, there are underwater fossils to be found in Florida's Withlacoochee Creek. If you'll settle for plants rather than animals, Boot Hill's Stonerose Interpretive Center (outside the little town of Republic, Washington) not only is rich in 100-million-year-old flowering plants but positively urges the visitors to dig for themselves. You pick up a permit and rent a geologist's hammer, and when you're through you're allowed to take a maximum of three of your finds home with you.

Well, I can't list every place in the world where you might find fossils. There is, however, one more that I'd like to mention, though it's not very convenient for most Americans. It is, in fact, a former limestone quarry in East Kirkton, Scotland, not far from Edinburgh, but it has two very special things going for it. For one thing, it has some of the oldest fossils around—338 million years old, to be exact—and for another, the local civic authorities have decided they have a major tourist attraction in the site and (as of this writing) they're getting ready to open it up for visitors.

Perhaps you don't want to travel that far? Fine. You can find a lot of fossils simply by opening your eyes as you make your daily rounds, doing what I call urban paleontology. And I'm sorry to say that, at a time when I was literally brushing up against fossils several times a week, I didn't know enough to do that.

I mentioned a moment ago that limestone is a good place to look for fossils. What I didn't mention is that when limestone is subjected to sufficient heat and pressure, it turns into that prettier stone, marble. Then some of that marble is quarried out, cut and polished, and used to decorate walls in the lobbies of many apartment and office buildings—the lobby, for example, of the building at 666 Fifth Avenue in New York.

Happens that the publishing firm of Bantam Books had its offices in that building for some years; happens, too, that I spent a lot of time there, as editor or contributor, and it never once occurred to me to take a closer look at that marble lobby until just before I resigned. And then I immediately perceived that those walls were just loaded with what was left of the mortal bodies of minute ancient creatures, smaller than a fingernail, captured in every position and orientation . . . and, until that moment, totally unnoticed by me.

The fossils these stones contain aren't anything as grand as dinosaurs, or even trilobites. They are much tinier creatures, but many of them are far older than any dinosaur, and they're not just in the walls of that one building on Fifth Avenue. Next time you're at your office job stop for a minute on your way to lunch and see what you can find in the lobby walls. If you take a good book on microfossils out of the public library, you may even be able to identify some of them and know that you are looking at, perhaps, foraminifera from the time when life on Earth had not yet left the sea.

Marble isn't the only construction stone that may be fossiliferous. That reddish sandstone that makes up so many of what are called brownstone-front apartment houses was typically laid down in the Triassic Period, some 200 million years ago. Major quarries of this rock are in the Connecticut River valley, and it has been used in century-old buildings all over the eastern United States. These stones don't contain many bones. They do from time to time, however, contain some of those dinosaur footprints I was talking about a bit ago. So if you happen to see any brownstone in a building—particularly if you see a brownstone front that is being torn down, exposing previously hidden faces of the rock—give it a look-over for footprints. If you find any, they will be three-toed and quite birdlike—but you won't mistake them for a bird's. These prints are much bigger than your average bird might leave, since some of them are more than two feet across.

One other thing before we leave the subject. If you happen to visit the J. Paul Getty art museum, perched on a ridge of the Santa Monica Mountains just outside of Los Angeles, you can look for fossils there. I know the Getty is an art museum. They had no intention of exhibiting fossils, but when they bought the limestone cladding from a quarry near Rome, Italy—mostly because it was cheap!—they turned out to get more than they expected. Look carefully at those panels and you'll find traces of twenty-thousand- to eighty-thousand-year-old leaves, branches, fish, and even a deer antler, and, hey, you've had the equivalent of two museum visits for the price of one.

ALTHOUGH HUNTING FOSSILS on your own is great fun, it must be said that if you want to see what any prehistoric animal actually looked like, the place to go is a museum.

When fossils are found they are almost never in good condition. Pieces are missing—sometimes almost all of the pieces are missing and there's nothing to see but a jawbone, a vertebra, or a tooth. Very often many of the pieces may be found, but almost inextricably mixed with assorted bones of other creatures, even of other species. (Canada's western provinces have huge bone beds, miles long, where an ancient river swept countless animal corpses into one great jumble, before covering them over.) Most of all, they are generally *squashed.* I had heard about the famous 10-million-year-old archaeopteryx long before I ever had the chance to see it at firsthand. It was as advertised. You could see the imprint of its primitive feathers, but the weight of layer after layer of rock above it had pressed it pancake flat.

So paleontologists spend a lot of time trying to fit bones together—like a three-dimensional jigsaw puzzle with many of the pieces missing—and to tease the bones out of the rock they are embedded in. It isn't a job for beginners. To convert those broken, jumbled bones into a recognizable shape requires hammers, chisels, dentists' drills, small sandblasters, acid, hardeners, various other chemicals . . . and, oh, yes, great skill. But you don't have to take my word for it. A few museums have begun to let the visitors see the process for themselves. The first I know of, or at least the first I happened to see for myself, was the London's great National Museum of Science and Industry in South Kensington, which had built a little glass-enclosed cubicle in the main hall, wherein restorers plied their trade. Then Chicago's Field Museum did it, to let the visitors watch the delicate years-long job of extricating their prize T. rex, "Sue," out of its rocky matrix. Others have followed.

Every natural history museum in the world has a fossil section. I haven't seen them all, but I have managed to get a look at a fair number. A few have been disappointing, for one reason or another—for instance, the natural history museum in Shanghai, China. I had the privilege of being shown through it by its assistant curator, a worldly man in Bermuda shorts, smoking Marlboros in a long cigarette holder. Because China has a long history and has preserved a lot of it, I expected a lot from this museum. I didn't get it. When the tour was over and the man asked me what I thought of his museum, I cleared my throat and said, as politely as possible, "You certainly have a lot of space, don't you?" He grinned, a little sadly. "What you mean," he said, "is that we don't have very many exhibits. We had a lot more once. But then every last thing in the museum, down to the showcases and the light fixtures and of course the exhibits themselves, was torn out and destroyed by the Red Guards in the Cultural Revolution and we haven't yet been able to replace them all."

But most museums are not disappointing. I particularly treasure the two I mentioned above, plus New York's American Museum of Natural History, with probably the world's best collection of dinosaurs and much, much else. Plus the Royal Tyrrell Museum in Drumhoolie, Alberta, Canada, northeast of Calgary and conveniently handy to one of those great bone beds I spoke of. Plus one other.

That other is a quite small museum—the whole thing would fit easily into the central hall of the large ones I've just mentioned. It isn't very convenient for most Americans to visit, either, but I like it a lot. It's the National Museum of Kenya in Nairobi. It doesn't have much in the way of dinosaurs, but it does have some handsome dioramas of early humans, lifelike reconstructions of gracile (means skinny) and robust (means not at all skinny) australopithecines posed on a nice reconstruction of the savannah they lived on a million or more years ago. Of course, a lot of museums have some of that sort of thing, but Nairobi also has something I've never seen anywhere else. That's a reconstruction of the dig at Kooba Fori where the Leakeys found some of their best prehuman fossils—almost as good as visiting that famous site yourself, and a lot less difficult.

But what Nairobi's museum has that makes it really special is that it is *there*. The ground the museum is built on may well be ground that some of those same protohumans trod long ago, when the world was young, and what other museum can say as much?

One other thing about museums. Many of them have the pleasant habit of arranging field trips. That's the best way to go fossil hunting, in the company of a professional paleontologist. Check with your nearest natural history museum, and wear your sturdiest shoes and a pair of slacks that you don't mind getting muddy.

APART FROM A couple of brief references to our distant relatives (but definitely not our ancestors), the australopithecines, I've been talking about the past as if the human race didn't exist in it at all. Of course this is untrue. Human beings are relatively young among Earth's species—the oldest definite indication of anatomically modern humans goes back no more than a hundred thousand years or so, contrasted with 100 million for an average dinosaur—but they certainly were there. Besides, they did something that other species almost never did! They didn't just leave us their bones, they also left us some of their works. They left us those wonderful cave drawings that we talked about in Chapter 7, and most of all they left us some spectacular ruins.

* * *

NOT MANY ARE more spectacular than England's Stonehenge, less than an hour's drive from London. If you don't have a car it's easy enough to get to anyhow; there are day excursions from London that take it in (though you may have to visit a couple of cathedrals along the way).

The first time I visited Stonehenge was in 1965, a time now referred to as the Good Old Days because you were still allowed to roam freely around the whole complex. I was particularly glad that that was so, because I had a quest. One of the central monoliths had just been reported to have the shape of a clearly Cretan dagger carved into it. That led to a flurry of excited speculation among anthropologists: Could it mean that the builders of Stonehenge, whoever they were, had been in some sort of contact with the great Minoan civilization on the island of Crete in the Aegean Sea, an ocean or two away? (I found the carved dagger easily enough, leading me to marvel that, over the many centuries those stones had stood there, no one had ever noticed it before.)

The Cretan theory seemed promising for a while, especially as there turned out to be a string of similar megalithic ruins in France and the Iberian peninsula. If you drew a map and connected the dots, it was easy to conjecture that some Cretan explorers had traveled west from some Mediterranean port, teaching the locals how to build great big stone structures as they went. Radio-dating did the theory in. It turned out that Stonehenge was the oldest of the megaliths, not the newest as it should properly have been if it were at the end of a Cretan cross-European tour. What's more, the dating showed that Stonehenge had been constructed a thousand years or more before Crete flourished.

Whoever built it, Stonehenge is an impressive sight: huge, carved rock pillars, made of a stone that can be found only many miles away and somehow, no one knows how, the builders managed to lug all the way to its site.

It's not as easy to roam around Stonehenge anymore. Blame that on the New Druids. They insisted on their religious right to have their Midsummer Sunrise ceremonies there every year, until the police got tired of the jeering crowds and kept everybody away on that date. Blame it also on visitors who liked the stones so well that they wanted to chip pieces off them to take home, which is why the central structures are now off limits. In fairness to today's tourists, I should say they didn't invent that particular vice. Indeed, a century or two ago merchants in the nearby towns made a good thing of selling chisels and hammers to would-be vandals.

Actually, Stonehenge never had anything to do with the Druids, of whatever vintage. Who built it, and exactly why, is still unclear. The theory that it was intended as a sort of early astronomical observatory seems plausible, but it tells us nothing about why those prehistoric Britons cared that much about the sky.

Still in Britain, there are burial mounds in plenty, and at Charlton, in Hampshire, there's an actual reconstruction of a prehistoric Iron Age community. It's called the Butzer Farm, and a working farm it is, farmed just as the early Britons did before the Romans came.

But the Romans did come, and left their mark all over Britain. They stitched the island together with fifteen hundred miles of their famous Roman roads, stretches of which still survive. (And other stretches of which are merely buried under contemporary highways, like London's Watling Street.) Not only are there countless cataloged finds of Roman buildings, but new ones are still being discovered. East of St. Paul's Cathedral in the city of London itself a previously unknown Roman temple turned up when ground was excavated for a new skyscraper. The temple's contents have been removed to museums (notably the Museum of the City of London, worth a visit on its own merits), but when it appeared the building's architect swiftly altered his plans. The two-thousand-year-old temple is now part of the building's foyer. There's a fine Roman villa near the Surrey-Sussex border, southeast of the city. Visitors can admire the ingenuity of Roman builders who, not caring for the damp and dreary English winters, invented central heating. You can see where open tiles circulated hot gases from a furnace underneath the flooring in the system called a hypocaust. (Leading my then–fourteen-year-old son, as he gazed on it, to make a joke: "What's a hypocaust?" "Oh, about five bucks, but you can get one for nothing at the needle exchange.") Of course, when the Romans pulled out they took the secret of central heating with them, not to be rediscovered in England until the twentieth century.

Biggest by far of the Roman ruins in Britain is Hadrian's Wall, meant to keep the fiercely battling Picts out of the civilized (i.e., Roman) lands to the south. It bisects the island from coast to coast, just below the Scottish border. Most of the wall is gone, but goodish stretches remain, along with the ruins of soldiers' barracks, and temples and stables, requiring not much of a detour if you are driving to, say, Edinburgh.

IMPRESSIVE AS HADRIAN'S Wall is, it is only as a suburban picket fence to the one the Chinese emperors built to keep out their own unruly northern neighbors. The Great Wall of China is certainly the largest man-made structure on Earth, running four thousand miles westward from the Yellow Sea to the steppes of Central Asia. It snakes up and down hills and valleys, wide enough to carry wagons on its top, with guard posts every few hundred feet. There's a bit of it, in good condition, only a short drive from Beijing. Getting there is easy enough, walking along it harder for sedentary Americans, because it is so steep—but if you do, then you can buy from one of the vendors a T-shirt that brags, in English and Chinese, "I have climbed the Great Wall." Actually, the vendors are good-hearted people, and they'll sell you one even if you haven't. The wall was first built four hundred years before Christ, built and rebuilt for nearly another two thousand years, all to keep the barbarians of the north out of the Middle Kingdom.

And, of course—like Hadrian's Wall, like the Maginot Line, like all the other static fortifications the human race has built—in the long run it failed.

WAY BACK WHEN I was talking about volcanoes I mentioned the island of Santorini, in the Aegean Sea, and its relation to the great empire of Crete. Crete itself is still there and well worth a visit for its great (if ruined) castles and artwork.

If you take the surviving Cretan art to be literally true, you have to suppose that King Minos and his subjects were anatomically pretty bizarre, since they are all depicted, male and female alike, as if they were as wasp-waisted and buxom as a Floradora girl. Probably that wasn't true. Probably it is just a convention of Cretan art—the same sort of convention that, for instance, made the Egyptians in the time of the pharaohs all show themselves with their heads on sideways.

Whatever they looked like, the Cretans were an enterprising lot. Their long, skinny, mountainous island stands at the lower end of the Aegean Sea. During the Bronze Age, sixteen or so centuries before Christ, it was home to the first real cities, and thus to the first true civilization, in Europe. The Cretans had it all. They had writing—a skill not common among other Europeans at the time—and proved it by leaving clay tablets for us to puzzle out. They built great palaces of stone and mud-brick, especially at Knossos, where the great king Minos watched his bull-dancers risk their lives for his amusement (in what is probably the sport that has come down to us as the bullfights of Spain and Mexico). Somewhere I have a photograph of myself sitting in Minos's very own regal throne in the amphitheater where the bull dances were held.

Of course, what most of us know about Minos is his bull-headed Minotaur and the labyrinth where the young victims were sacrificed to the monster until Theseus came along and killed it off. How much of that is true is arguable. What isn't arguable, though, is that the Minoans made at least one wonderful invention that died with their empire and was not reborn for thousands of years.

As you explore the queen's chambers in the ruins of the palace at Knossos, you can see it for yourself. It is a stone seat with a hole in the middle of it, over what was then a gently flowing stream, and you are in the presence of history's first flush toilet. (But you won't be allowed to get close to it anymore. The stream no longer flows, and the authorities have roped the chamber off—apparently because some tourists did, now and then, try to put it to its original use.)

That first-of-all European civilization didn't last. Invaders from Mycenae, on the Greek mainland, took them over, perhaps helped by the destruction caused by Thera's eruption. From then on the center of power in that part of the world was in Greece itself.

You don't have to be told that there are some wonderful remnants of the early Greeks. Certainly no visitor to Athens can avoid seeing the Parthenon majestically perched on its hill overlooking the city.

On closer inspection, perhaps it's not quite as majestic. During a war a century or so ago the Greeks stored ammunition in it; naturally the enemy shelled it, naturally the ammunition blew up. Earlier still, the English Lord Elgin admired its marble friezes so much that he had his workmen rip them out and ship them off to London—which is why the British Museum is one of the best places to see what ancient Athenian art is like. This does not please the Greeks, who have complained about it for generations, though, actually, when you consider what the corrosive Athenian air has done to most of the remaining friezes, Lord Elgin may have done them a service. In spite of all of this, the Parthenon is a noble sight, and, if you don't happen to be planning a Greek vacation in the near future, there's a full-scale reproduction of it, in considerably better repair, in Nashville, Tennessee.

There are plenty of Greek ruins that aren't actually in Greece, anyway, from Alexandria in Egypt (named after guess who?) to Naples (called Neapolis, or "New City," by its original Greek settlers) in Italy. Sicily in particular is full of "Greek" theaters. (Sure, they were originally built by the ancient Greeks, but then they were so thoroughly "modernized" by the Romans that Aeschylus and Euripides might not recognize them.)

There are a couple of museums in Sicily. The big one in Syracuse, called the Museo Archeologica Regionale "Paolo Orsi," has some nice Greek art and anyway it's worth a look on its own merits for its odd, high-tech, daisy-shape design. One that I like is in the little town of Naxos, nearby. Naxos was named for the Greek island its original settlers came from, in 735 B.C. The interesting thing about this little museum is that it covers the complete history of Naxos. That history was shortened when it took the wrong side in the war between distant Athens and nearby Syracuse; the reigning tyrant of Athens, Dionysius I, had no trouble conquering Naxos, and taught the people a lesson by razing the city and enslaving them. That was nasty for the Naxians, but it was history's gain. Naxos never again amounted to much as a town, which means that its ruins weren't built over, and its whole life span is available to the diggers.

THE CITY OF Rome, on the other hand, never stopped being a great city, which makes it all the more remarkable that so much of the Rome of the Caesars is still visible. There's only one ancient Roman building that is still in daily use—Marcus Agrippa's Pantheon, now converted to a Catholic church—but you can't walk five blocks on any of the seven hills of Rome without treading on history. The Roman Forum still exists. At least some of it does, though it's in pretty bad shape. Still, you can mark where the senators guided the affairs of most of the known world. You can even find the spot where Julius Caesar was stabbed to death by his colleagues, now a rather undistinguished bar. The Colosseum still stands, too, though in almost as bad shape as the Forum, due to the habit of intervening generations of burning its marble to make quicklime and pulling apart its great stone blocks to salvage the metal pegs that held them together.

The trouble with Rome as a source of Roman history is that it is scattered and disconnected. How much nicer to find a fossil city that was kept untouched for all those two millennia, with everything still where it was.

Such a place does exist. It's called Pompeii.

Pompeii is special. Over the long and frequently violent history of the human race, many cities have been destroyed one way or another, by natural catastrophe if not by war. In the case of Pompeii, the natural catastrophe that destroyed it also preserved it.

A few chapters back I talked about the volcano, Vesuvius, looming across the bay from Naples, Italy. It has erupted many times in history, but never more violently than on the day in A.D. 79 when its burning ash buried the pleasant little Roman city in an afternoon, killing everyone who lived there.

Because it was over so quickly, what it has left is a sort of snapshot of how life was lived two thousand years ago. How did they do their laundry? Soiled garments were soaked in a mixture of water, lime, and human urine—a row of jars stood outside the laundry and passersby were urged to relieve themselves there to provide the laundrymen with raw materials—and then slaves trampled them with their bare feet. Baking? It's all there in the bakery on the Vicolo Torto: grain milled with the hardest stone available, so only a minimum of stone flakes might get into the flour and chip a Roman tooth. Baths? Three of them in Pompeii, with their hot tubs, warm tubs, and cold plunges, with a swimming pool at the end. Shops and workshops? Stroll down the Via dell'Abbondanza and you will pass a smithy, a felt-maker, several quick-food and quick-drink shops—a cross between McDonald's and your corner bar, well adapted for loafing—and an inn that once belonged to a man called Solericus. His slave waitresses served drinks on the ground floor and themselves one flight up, for Pompeii was not lacking in brothels. There were temples, forums, private homes for the well-to-do, the latter lavishly furnished with gardens and murals and works of art.

I don't want to make Pompeii sound like more than it is. It's not a theme park. Its facilities are no longer in working order. There is no wine in the jars of the fast-drink emporia, and even the jars themselves have generally been taken to some museum. There are no bodies, either. The ones that were not simply annihilated in the fall of hot ash have long since rotted away.

But not without leaving a trace. Remember what I said about the fossils that exist only because the rock around them has preserved their shape? So, too, in Pompeii. There's a man caught in such detail that the straps of his sandals can be made out, a mule-driver and his mule, a dog caught in a final howl of terror.

Pompeii is special. Other ancient cities have been excavated, but only fragments were left for the archeologists. There's nothing like Pompeii.

Well, almost nothing.

To be precise, there is at least one other spot that comes close to Pompeii in its pristine integrity, and it happens to be in our own hemisphere.

PRE-COLUMBIAN AMERICA HAS left many monumental traces of its lost civilizations. There are the Cahokia Mounds, in my own home state of Illinois, and the abandoned cave dwellings of the Anasazi, near the Four Corners of the southwestern United States. When Mexico City dug its subway it turned up magnificent works, now preserved for commuters to view at their station. I remember sitting in the shade at Altun Ha, the Maya ruins in Belize, while my wife and a couple of other visitors more energetic than I climbed one of those Mayan pyramids, so like those of Egypt. (Not because Egyptian tourists came by and left off the blueprints, as some would like to think. After all, if you want to build something really tall and have nothing but cut stone to work with, a pyramid shape is what you pretty inevitably wind up with.)

But of them all, the most wondrous is Machu Picchu, in Peru. Alas, it's not easy to get to. But it's worth the trip. You start by flying into the modern mountaintop city of Cuzco, and if you are wise you stay there for a few days. There's plenty to see, because Cuzco was the capital of the Inca Empire before Pizarro came along and changed everything, and anyway you may want to get used to the thin air at the city's twelve-thousand-foot altitude before taking the train to Machu Picchu.

Machu Picchu is not a demolished city. It is simply a deserted one. It is a great fortress hidden in some of the most desolate mountains in Peru—hidden so well that the Spaniards never found it, and it was lost and forgotten until an American archeologist named Hiram Bingham came across it in 1911. Although it has been studied intensively, it remains a mystery. No one is sure just when it was built, or even why; the best guess is that it was built by Inca rulers as a place to put their families for safekeeping during some now-forgotten war. But then it seems that its occupants, whoever they were, simply walked away and left it.

But there it is, a complete five-square-mile city on top of one of the least accessible mountains imaginable. The labor it must have taken to build it is staggering. Visitors to the Egyptian pyramids marvel at the fifteen-ton stone blocks that fit together so tightly that you can't slip a knife blade between them. Machu Picchu's blocks are as large, and they do not fit at plane faces, like bricks in a wall. A block may have jogs and depressions that are faithfully matched in reverse in the block it fits against. Why such unnecessarily difficult construction? No one knows.

There was another, much lesser mystery that confronted us as my wife and I rode the car back down to the train in the valley below. The mountain is so steep that the road up to the old city is a series of long, repetitive switchbacks, and we noticed that at every single hairpin right turn (no guard rails, nothing but steep slopes to roll down if our driver missed a curve) a small boy stood and waved to us. There was none at the left-hand turns at the other ends of the switchbacks, and we wondered where the locals found a dozen identical-looking boys for so trivial a task.

That mystery, at least, we did solve. It had to be the same one boy at each turn, scrambling down the slopes while we took the long switchbacks. Would that Machu Picchu's other puzzles were as easy to solve.

Chapter 9
SHOPTALK

Meetings and Lectures

Although it's fine to see science of some kind actually happening, that's not the only pleasure to be found from chasing science. It's also fine to listen to the scientists themselves when they're talking shop.

One way to hear them is to show up when they give a public lecture, which even the best of them do from time to time for lay audiences. Alternatively, you can go so far as to attend some of their professional meetings at symposia and conferences. Over the decades I've been to a mort of the things. Some were better than others, but there weren't any that I wouldn't gladly do all over again. It is true that some of the ones I remember with most pleasure were invitation-only affairs, but it is a fortunate fact that some of the very best scientific meetings around are open to anyone who cares to show up at the meeting hall and pay the registration fee.

At the very top of the list of generally accessible all-sciences conferences I would put the annual February meetings of the American Association for the Advancement of Science—shortened as the AAAS and usually pronounced the Triple-A S. These are a movable feast, wandering from city to city each year. Recent ones have been in Boston, Chicago, Washington, and San Francisco, and they are almost sure to turn up at some city fairly near you sometime in the nearish future. It has to be a pretty big city, though, because there will be several thousand people at each one and the conference needs a fairly lavish amount of hotel and meeting space.

The AAAS meeting lasts just about a week each year. It has two main purposes: to give scientists a chance to see what else is going on in their fields and others', and to let the public listen in. The audiences are about fifty-fifty of each category.

In either category, what you get when you go there is fine. I've attended AAAS meetings in all sorts of capacities—as program participant, as press representative, and as a person with no official status at all, like anyone else who might happen to wander in off the street. The difference is small. I admit it's nice to have the privilege of the participants' lounge and the press room, but that's only a little icing on the cake. It's what goes on at the meetings that is important, and I've had a grand time at every one.

How could I not? How could I not enjoy hearing a scientist tell the inside story of how he set up seismic nuclear-test monitoring stations in the old Soviet Union, and what he found out about them? Or hearing astronomers from the Harvard-Smithsonian explain how they had analyzed red shifts and sky maps in order to locate every galaxy in a sector of the sky, giving each its proper direction and distance, programmed the data into a computer and instructed it to draw a three-dimensional picture of that slice of the universe, several thousand light-years across . . . and then have them go on to show us that picture and even rotate it for us so we could see a God's-eye view of the universe from *outside*? Or listening to a physicist from the National Oceanic and Atmospheric Agency tell us, with slides and diagrams, what she had learned on an expedition to study the Antarctic ozone hole, or hearing an archaeologist describe the real puzzles of the Egyptian pyramids, or a pharmacologist tell about the herbs and leaves the monkeys of the Amazon jungle eat, not as food but, apparently, for their medicinal value—the same plants the natives harvest for the same purpose?

I could go on indefinitely, just listing the treasures in an AAAS meeting, without covering them all. As a matter of fact, it is impossible for me to cover them all, because I have not attended them all. If the AAAS meetings have a fault, it is overkill. There are generally six or eight symposium tracks going on at a time, plus tutorials, plus plenary lectures, plus poster exhibits; the big problem isn't in finding something worth attending, it's in making those painful decisions about which promising performance has to be missed in order to sit in on an even more promising one. You can't do everything at an AAAS meeting, but the fraction that you can do is golden.

What's more, there is always the chance to strike up a conversation with some like-minded person at such a gathering—at lunch, around the coffee urn during program breaks, taking a five-minute breather in a hotel lobby. Even your vices can work for you. If you are a smoker, as I was for many years, you join the fraternity of addicts as they huddle in whatever ghetto is allowed them, in a sealed room with glass walls or under a hotel marquee to avoid the rain. Conversations flourish. Or you can simply eavesdrop, perhaps on someone who's busily radiographing Native American skulls (before being forced to return them to their descendants for reburial), while that person is chatting with someone who is computer-modeling the climates of the Cretaceous Period.

THE AAAS MEETINGS try to cover all of science, and come pretty close to succeeding. But there are plenty of other, more specialized scientific meetings going on all the time, great and small. There is no way to list them all, especially since some of the best are one-time, ad hoc gatherings convened for a special purpose and not repeated.

Those special-interest meetings are often more narrowly technical than we would most enjoy, but even the most specialized have their rewards. I did not, for instance, expect a lot from the conference on public utility energy storage I attended some years ago. My main reason for attending it was the fact that it was held in beautiful Dubrovnik, on the Adriatic coast of what was then Yugoslavia (and has now, sadly, become a multi-national battleground). But there were more ways of storing energy than I had expected—pressurized gas, water pumped uphill, batteries, spinning high-density wheels, and a dozen others—and I came away with a new understanding of problems I had not even known to exist.

What I remember best about one seminar on planetology and space-mission planning, held at the Waldorf-Astoria in New York, isn't the good papers or the exciting prospects that were discussed there. It's one particular scientist who had made the bad decision to use that forum to announce his revolutionary new General Theory of Everything, a scheme that took up where Einstein and Planck left off.

It did not go well for him. Most of the audience sat politely stone-faced, but at the rear of the hall an Italian astronomer from the Institute of Advanced Studies in Princeton was pacing back and forth, muttering to himself in Italian. My command of the language is limited, but I could catch the main thrust of his mumbles; subtracting the expletives, what he was saying was "But this is the most preposterous insanity!" I was glad to hear him say it; I had had that same suspicion for myself.

A number of specialized scientific societies run their own annual meetings open to the public, and they can be fun. On space travel and related themes, the American Astronautical Society (the *Double-*

AS, that is) holds its Goddard Memorial Symposium every year in Washington, and all are welcome to attend. Alternatively, there is a similar annual event run by the AIAA (or American Institute for Aeronautics and Astronautics). They do not quite duplicate each other, either. AAS's Goddard series tends to be less technical and more far-ranging; AIAA's meetings are more concerned with hardware and with practical applications. Until recently, at least, the AIAA's emphasis was heavily on the military uses of space. The last AIAA meeting I spoke at had more generals in attendance than I had ever seen in one place since I was based at the headquarters of the Mediterranean Theater of Operations Air Force in World War II, and the displays in the exhibit hall concentrated on missiles and military surveillance.

Another good annual meeting open to the public is the one held by the American Association for Artificial Intelligence (or AAAI . . . which is not to be confused with the AIAA above; does it seem to you that we're running out of initials?). The AAAI is where the country's robot-builders and fuzzy-logic experts get together to compare notes each year. It moves around from year to year, and the plenary sessions at an AAAI meeting are fine. So are the demonstrations of recent projects, though some of the more technical papers may be most interesting to the specialists. They may be a little disheartening, too. During one AAAI papers session, I happened to sit next to a man from a classified Department of Defense research facility. As the nonclassified researchers presented their papers on their latest findings in robots and computers, every few minutes he would lean over and whisper to me, "We did that five years ago," or "Sure, we tried that, but it doesn't work."

APART FROM THE purely scientific meetings, there are others that sometimes have interesting science-based sessions. The World Future Society's main interest is, as you can easily enough guess, the future in all its ramifications: forecasting methodologies, anticipations of future economic and social events, analysis of current trends, and so on. Its general meetings are the usual multitrack events, but among the tracks there are frequently interesting sessions on scientific topics. There are, too, at such perhaps unpromising sites as the Reverend Sun Myung Moon's annual Conferences on the Unity of the Sciences. I hesitate even to mention these—friends have given me a hard time for attending the one or two I've spoken at—but it is at them that I've met such people as Paolo Solari, the man who builds "arcology" habitats in the Arizona desert, and even an occasional Nobel laureate. And I would be remiss if I didn't at least mention the opportunities for chasing science on my own native turf, the world of science fiction.

If the whole idea of science fiction turns you off, you can move right on to the next page. I'm not trying to convert you. But many, though not all, science fiction readers (and writers) are as avid in the chase after science as anyone else, and that is often reflected in their "cons."

What, you may ask, is a "con"? That is the science fiction fan-speak term for any gathering of people who are interested in science fiction, and science fiction cons have become big business. Every year there is the "worldcon," generally held over Labor Day weekend and usually in a large city in the United States (but occasionally in some other, generally English-speaking country). There are also hundreds, literally hundreds, of local and regional cons all over the world every year. Their main thrust is in the books and films of science fiction—often in fantasy, as well, sometimes in comics or game-playing, too—but many of them also include some kind of science in their programming. There will almost always be some program items on space travel, usually with people from the National Space Society in attendance, now and then with an astronaut. But that doesn't end it. The annual Boston-area cons (they are called Boskones, for reasons that SF fans probably will recognize and are too complicated to explain to anyone else) make good use of such local resources as Harvard, Route 128, and MIT; it was at a Boskone that I first met Warren McCullough, one of the world's great authorities on the relationship between intelligence and the architecture of the human brain, as well as a fair number of computer mavens, astronomers, mathematicians, and nuclear physicists who shared a fondness for science fiction with their professional interests. At a Toronto con a pair of aeronautical engineers showed off the working ornithopter they had just succeeded in making fly; the Chicago events draw on people from Fermilab, Argonne, and the universities; West Coast events are likely to have someone from the aerospace industries on the program. Sometimes the scientists are pretty celebrated. In my own case, the one scientist I most wanted to meet was Richard Feynman; a few years ago I went out of my way to attend a con in Minneapolis largely because he was scheduled to be there. (In any ordinary year I might well have attended anyway, because it's a good annual con, but that one wasn't convenient since I happened to be living in England at the time.) But I missed meeting Feynman there, or anywhere, because, sadly, he died shortly before the con.

IF THERE DOESN'T happen to be a good scientific convention going on within visiting range, there may be a lecture series sponsored by some local institution.

They are presented all over the country. Three that I particularly like are the ones run by Fermilab in the Chicago area; the Boston series run jointly by the Harvard-Smithsonian and the Boston Museum of Science (at the museum); and those sponsored by the New York Academy of Sciences. Some of those last have been particularly great, especially one given by P. A. M. Dirac, discussing his large number hypothesis. (What is the large number hypothesis? Don't ask. Or better still, do ask the next top-ranked cosmologist you happen to be having a drink with, and be prepared to be told either that it's utter rot or that it is one of the most important observations ever made.)

The New York Academy lectures do have some unfortunate aspects. The academy building, just off the Fifth Avenue border of Central Park, has less auditorium space than is really needed when a celebrity like Paul Dirac comes to speak, especially when every college physics teacher in the city has told his or her students that they'll get class credit for attending. But in the regular course of lectures the audiences fit cozily into the available rooms, and the wide-ranging talks can be great fun.

In the Chicago area, among the best lectures are those at Fermilab (down in Batavia, Illinois; we talked about it at some length in Chapter 1). The Fermilab lectures are not always easy for a layman to follow. I recall one on string theory (don't ask about that one, either). Halfway through I muttered in my wife's ear, "I'm understanding about one word in three of this," to which she—the college professor with the doctoral degree—replied, "And *I'm* understanding about one in ten." But not all the lecture subjects are that hairy, and anyway a good lecturer usually can make almost any subject comprehensible.

The best lecture on a tough subject I ever heard at Fermilab was given by Stephen Hawking, on black holes. Hawking was brilliantly lucid in discussing this pretty arcane subject, but what made things particularly tricky was that the lecture had to be given under Hawking's enormous physical handicaps. Hawking's ALS (amyotrophic lateral sclerosis, or Lou Gehrig's disease) had long since wasted his body to near-total paralysis, and not long before the lecture it had taken away his voice, as well. He could talk to us only through a voice-synthesizing computer.

The bulk of Hawking's lecture had been preprogrammed to save time, so that all Hawking needed to do was to switch the machine on—apologizing as he did so for the computer's American accent; the device had been built for him in California. It was not until it came to the time for questions that things got difficult. It took Hawking a good five minutes, using the few of his fingers that still worked on the keypad strapped to the arm of his wheelchair, to generate a twenty-second answer to a question. The answers were, of course, delightful. But after half an hour of this Hawking's moderator, a Fermilab scientist, cleared his throat and spoke up. "I would urge the audience," he said politely, "to bear in mind that there are a great many quite fundamental and far-reaching questions that can be answered with a simple 'yes' or 'no.' "

Whereupon the very next questioner piped up with "Dr. Hawking, please tell us how the universe was formed in the Big Bang."

YOU DON'T HAVE to live near Fermilab or the New York Academy of Sciences to find good science lectures, though. Most schools, from community colleges to universities, have continuing lecture programs open to the public. So do many libraries and museums. All you need to do is find out about them. Newspapers often publish lists of such events, particularly in smaller communities; failing that, they may be listed on the bulletin boards in local libraries and schools.

Or you can take matters into your own hands. Call up the nearest museum, planetarium, research facility, scientific organization, or university. Ask them (a) if they have a science-oriented lecture series for the layman and, if so, (b) if they will be kind enough to send you a schedule. Most will be glad to do so; after all, they *want* people to attend their lectures.

Chapter 10
SCIENCE BY MAIL

Publications, Organizations, Plus TV and the Web

When all else fails, there is always one good way to chase science in your own home. That is the ultimate resource of the literate human being: reading.

Probably there is more good reading on science available today than ever before in the history of the human race. Books like Stephen Hawking's *A Brief History of Time* even sometimes make the best-seller lists, and there are literally hundreds of other excellent works on the market. (Well, unfortunately there are just as many junky ones, too. Remember the old Roman adage, *caveat emptor*, which may be approximately translated as "Taste the soup before you swallow it.")

Some of my own personal favorite books are Hans Moravec's *Mind Children*, on the future of computers; Hans Christian von Baeyer's *Taming the Atom*, on nuclear physics; and almost everything by John Gribbin, Marcus Chown, Carl Sagan, Isaac Asimov, or Richard Feynman.

There are also several dozen science-oriented periodicals, including some very good ones. What I would call the basic library of general-purpose science-oriented periodicals includes three weeklies—*Science*, *Nature*, and *New Scientist*—and around a dozen monthlies aimed at audiences of varying levels of scientific sophistication.

The grande dame of the monthlies is *Scientific American*, solid, trustworthy, and ecumenical. No area of science is foreign to *Scientific American*, and, if it isn't exactly written for a scientifically uninformed audience, it is at least relatively free of equations and arcane vocabulary. You may find unfamiliar terms in a *Scientific American* article, but they will almost always be explained as you go along; the thrust of the articles is not only to tell you what is interesting in science but to help you understand it.

A bit less hospitable are the two major weeklies, *Science* and *Nature. Science* is American, published by the AAAS (and comes with membership); *Nature* is British and unaffiliated. Both operate at the cutting edge of science, and their prestige is great. It is the dream of every scientist to see his newest paper in one of those weeklies, and so most of the more revolutionary developments in scientific research are first published in one or the other of them. Like *Scientific American* they are ecumenical; unlike it, most of the reports they publish assume a certain basic knowledge of the field. Probably the best description of the reader *Science* and *Nature* aim at is a person currently employed in one specific discipline of science who wants to know what is going on in the rest of the scientific world. However, both weeklies do make an effort to help the lay reader along. If there is a densely technical report on some complex, but important, subject at the back of the magazine, there will usually also be a nontechnical article explaining what the report is all about in the front.

The all-science weekly I personally happen to like best, however, is England's *New Scientist*. (I may be not without bias here, since I've done a little writing for it myself from time to time.) *New Scientist* does have its downside for American readers. For them it is fairly expensive, since it has to be airmailed across the ocean. And it is, after all, definitely British. This means there is a lot of space devoted to subjects of less than compelling interest to us colonials, like how many British academics are likely to lose their jobs over the next few months due to budgetary downsizing; and sometimes the humor in its cartoons is opaque to Americans. But it does have such reader-friendly things as cartoons and plenty of pictures along with the text. The whole magazine is written in a sprightly style; and *New Scientist* assumes *nothing* about its readers' knowledge of science.

Wonderfully, it manages to do all this without simultaneously becoming trivial. It covers everything, because its staff reads all the journals as well as commissions its own original material. What's more, simply because it is a weekly it does so fast. If I had to give up all my scientific periodicals but one—God forbid!—*New Scientist* is probably the one I would keep.

Beyond these basic publications there are *The Sciences*, published by the New York Academy of Sciences, and *Discover*, the only survivor of the popularized science magazines that bloomed in the 1970s and died the 1980s; *Discover* has gone through various ownership and editorial staff changes but seems to have settled down to being a reader-friendly magazine that is the easiest reading for the newcomer to science. There are specialized astronomy magazines (*Sky and Telescope*, *Mercury*—this latter comes with membership in the Astronomical Society of the Pacific—et al.), plus the publications of the British Interplanetary Society, *Spaceflight* and the *Journal*; there are ones that are not exactly scientific but do frequently have articles on scientific subjects that are worth reading, like *Smithsonian* and *Natural History* and even *National Geographic* (all three of which come with membership in, respectively, the Smithsonian Institution, the American Museum of Natural History in New York, and the National Geographic Society) as well as the various environmental magazines, such as *Sierra*.

For that matter, even such all-purpose magazines as *The Atlantic* and *The New Yorker* have from time to time published exceptionally fine science-oriented pieces. (Though recently *The New Yorker* has gone through several editorial changes and sometimes now seems to be less interested in science than in glitz.)

On related subjects, there are *The Bulletin of the Atomic Scientists*, the *Defense Monitor*, and *The Skeptic*—the first two are your best sources on what is happening in military-related subjects that the Pentagon doesn't want to tell you, the other specializes in debunking claims of ESP, reincarnation, alien abductions, and so on—as well as *The Futurist*, published by the World Future Society and (as you might guess) devoted to speculations about future development and attempts at finding ways to predict them.

There are, in short, dozens of periodicals available to slake the thirst for science. It is true that most of them are unlikely to turn up on your nearest 7-Eleven or other convenience store magazine rack, so if you want them, you may have to subscribe.

But before you go that far, you might like to get an idea of what you're getting into. There's a simple way to do that: Browse the periodicals in your neighborhood public library and see which ones catch your fancy. If they don't have at least a fair number of the ones I've mentioned, write a snappish letter to your local library board.

THERE IS EVEN a good assortment of scientific entertainment on the boob tube. It isn't always easy to find. On the major television networks the science-related shows are unhappily drowned out by the mass of game shows, cop shows, sitcoms, and soaps, but the PBS stations and such cable channels as Discover and The Learning Channel frequently have shows worth watching. (Sadly, the TV channels have a bad habit on occasion of blending a little bit of science with a lot of superstition—pyramidology, flying saucers, psychics, and so on—so be choosy).

Even that doesn't end it. If you are a computer networking person there are many electronic meeting places on the Internet for the science-minded. With access to the right ones, you may be able to sit in on actual discussions among scientists working on the same subjects, though perhaps thousands of miles apart. And there are Web sites for almost any scientific subject you care to name.

It's all there. If you can't chase science as much as you'd like in the flesh, don't despair. Through books, magazines, the Net, and television you can have quite a lot of it come right to your home.

Appendix
SCIENCE EVERYWHERE

Geographical Index of Places Mentioned in the Text, Plus Science and Technology Centers All Over the World

I haven't anywhere near exhausted the places where interesting science can be found. I didn't say anything about that whimsically strange midwestern construction called the Peorrery. You are aware, no doubt, that the object called an "orrery" is a desktop model of the solar system, usually representing the planets with marbles of various sizes wired at appropriate distances from a large central globe representing the Sun. The Peorrery is the same thing, but on a much larger scale. It is the brainchild of the director of the local planetarium in Peoria, Illinois; he took the planetarium's dome as representing the Sun, and then went out to locate the Sun's planets around the state of Illinois at distances to scale. Mercury is a marble an inch and a half in diameter in a school supply store in Peoria's North University Street; Venus almost four inches across and still in Peoria, in a State Farm insurance office on North Sheridan Street . . . all the way to Neptune (a fifteen-inch globe in a Chrysler dealer's showroom in Roanoke) and Pluto (a one-incher, in a furniture store in Kewanee).

Then there's Memphis's Mud Island, the long, narrow mud spit in the Mississippi River just outside of the city of Memphis. Mud Island isn't only *in* the Mississippi, the good folks at Memphis have placed the entire Mississippi River (or at least a scale model of it) *on* Mud Island. Once you walk across the bridge from Memphis you can wander along the model, studying its flows, observing all its curves and riverside towns and tributaries with a godlike view that Mark Twain might have dreamed of. And then you can brag to all your friends that you've walked the whole length of the Mississippi, all the way to the delta, and you did it in a single pass, without stopping to eat or sleep.

Most of all, I haven't done justice to the very houses of science themselves: the natural history museums, the planetaria, the aquaria, the zoos—and, above all, the museums of science. There are more than four hundred of these in the world. Every one is worth a visit . . . and to help you find your way to them, I append a list of all those in America and the world.

Enjoy!

UNITED STATES

Alabama

THE ALABAMA AVIATION HALL OF FAME
4343 73rd Street North
Birmingham, AL 35206

ALABAMA SCIENCE CENTER /THE WATER COURSE PROJECTS OF THE ALABAMA POWER FOUNDATION
P.O. Box 160
Montgomery, AL 36101

ANNISTON MUSEUM OF NATURAL HISTORY
800 Museum Drive
Anniston, AL 36201

CENTER FOR CULTURAL ARTS
501 Broad Street
Gadsden, AL 35901

THE GULF COAST EXPLOREUM
65 Government Street
Mobile, AL 36602

McWANE CENTER
200 19th Street N
Birmingham, AL 35203

SCI-QUEST
102-B Wynn Drive
Huntsville, AL 35805

U.S. Space & Rocket Center (p. 77)
One Tranquility Base
Huntsville, AL 35805

Alaska

Glacier (p.143)
Skagway, AK

The Imaginarium
737 West Fifth Avenue, #G
Anchorage, AK 99501

Arizona

Arizona Science Center
600 East Washington Street
Phoenix, AZ 85004

Biosphere 2 Center
32540 South Biosphere Road/P.O. Box 689
Oracle, AZ 85623

Flandrau Science Center & Planetarium
University of Arizona
Tucson, AZ 85721

Halle Heart Center—American Heart Association
2929 South 48th Street
Tempe, AZ 85282

Lowell Observatory
1400 West Mars Hill Road
Flagstaff, AZ 86001

Meteor Crater (p. 42)
Barringer, AZ

Whipple Observatory (p. 49)
Tucson, AZ

Arkansas

MID-AMERICA SCIENCE MUSEUM
500 Mid-America Boulevard
Hot Springs, AR 71913

MUSEUM OF DISCOVERY (FORMERLY ARKANSAS MUSEUM OF SCIENCE AND HISTORY)
500 East Markham, Suite 150
Little Rock, AR 72201

California

BAY AREA DISCOVERY MUSEUM
Fort Baker
557 McReynolds Road
Sausalito, CA 94965

BIRCH AQUARIUM AT SCRIPPS, SIO, UCSD
9500 Gilman Drive
La Jolla, CA 92093

CALIFORNIA ACADEMY OF SCIENCES
Golden Gate Park
San Francisco, CA 94118-4499

CALIFORNIA SCIENCE CENTER
700 State Drive
Los Angeles, CA 90037

CHABOT OBSERVATORY & SCIENCE CENTER
4917 Mountain Boulevard
Oakland, CA 94619

CHILDREN'S DISCOVERY MUSEUM OF SAN JOSE
180 Woz Way
San Jose, CA 95110

COYOTE POINT MUSEUM FOR ENVIRONMENTAL EDUCATION
1651 Coyote Point Drive
San Mateo, CA 94401-1097

Discovery Science Center—Launch Pad
3333 Bear Street
Costa Mesa, CA 92626

The Exploratorium
3601 Lyon Street
San Francisco, CA 94123

Explorit Science Center
3141 Fifth Street
Davis, CA 95616

Fresno Metropolitan Museum of Art, History, and Science
1515 Van Ness Avenue
Fresno, CA 93721

J. Paul Getty Museum (p. 164)
Malibu, CA

Hall of Health
2230 Shattuck Avenue (Lower Level)
Berkeley, CA 94704

Humboldt State University Natural History Museum
1315 G Street
Arcata, CA 95521

Huntington Botanical Gardens
1151 Oxford Road
San Marino, CA 91108

Jet Propulsion Laboratory (p. 87)
4800 Oak Grove Drive
Pasadena, CA 91109

La Brea Tar Pits (p. 156)
Los Angeles, CA

Lawrence Hall of Science
Centennial Drive

University of California
Berkeley, CA 94720

LAWRENCE LIVERMORE NATIONAL LABORATORY
East Gate, Greenville Road
Livermore, CA 94550

LICK OBSERVATORY (p. 51)
Pasadena, CA

LINDSAY WILDLIFE MUSEUM
1931 First Avenue
Walnut Creek, CA 94596

MARINA DISTRICT (p. 93)
San Francisco, CA

MONTEREY BAY AQUARIUM
886 Cannery Row
Monterey, CA 93940

MOUNT PALOMAR OBSERVATORY (p. 52)
Mount Palomar, CA

REDWOODS
northern California

REUBEN H. FLEET SCIENCE CENTER
1875 El Prado Way
San Diego, CA 92101

SACRAMENTO MUSEUM OF HISTORY, SCIENCE, AND TECHNOLOGY
101 I Street
Sacramento, CA 95814

SAN FRANCISCO ZOOLOGICAL GARDENS
1 Zoo Road
San Francisco, CA 94132

SANTA BARBARA BOTANIC GARDEN
1212 Mission Canyon Road
Santa Barbara, CA 93105

Santa Barbara Museum of Natural History
2559 Puesta del Sol Road
Santa Barbara, CA 93105

The Tech Museum of Innovation
145 West San Carlos Street
San Jose, CA 95113

Turtle Bay Museums and Arboretum on the River
Carter House Natural Science Museum
4800 Quartz Hill Road
Redding, CA 96003

U.S. Army Corps of Engineers—Bay Model Visitor Center
2100 Bridgeway
Sausalito, CA 94965

Ventura County Discovery Center
2100 East Thousand Oaks Boulevard
Thousand Oaks, CA 91362

Mount Wilson Observatory (p. 52)
Mount Wilson, CA.

Colorado

Children's Museum of Denver, Inc.
2121 Children's Museum Drive
Denver, CO 80211

Denver Museum of Natural History
2001 Colorado Boulevard
Denver, CO 80205

Discovery Center Science Museum
703 East Prospect Road
Fort Collins, CO 80525

Natural Renewable Energy Laboratory Visitors Center
15013 Denver West Parkway
Golden, CO 80401

Connecticut

DISCOVERY MUSEUM, INC
4450 Park Avenue
Bridgeport, CT 06604

DINOSAUR STATE PARK (p. 161)
Rocky Hill, CT

THE MARITIME AQUARIUM AT NORWALK
10 North Water Street
Norwalk, CT 06854

MYSTIC MARINELIFE AQUARIUM/INSTITUTE FOR EXPLORATION
55 Coogan Boulevard
Mystic, CT 06355

SCIENCE CENTER OF CONNECTICUT, INC
950 Trout Brook Drive
West Hartford, CT 06119

SCIENCE CENTER OF EASTERN CONNECTICUT
33 Gallows Lane
New London, CT 06320

YALE PEABODY MUSEUM OF NATURAL HISTORY
170 Whitney Avenue—Box 208118
New Haven, CT 06511

Delaware

HAGLEY MUSEUM AND LIBRARY
2948 Buck Road East
Wilmington, DE 19807

District of Columbia

CAPITAL CHILDREN'S MUSEUM
800 3rd Street NE
Washington, DC 20002

EXPLORES HALL, NATIONAL GEOGRAPHIC SOCIETY
1145 17th Street NW
Washington, DC 20036

NATIONAL AIR AND SPACE MUSEUM, SMITHSONIAN INSTITUTION
6th Street and Independence Avenue SW
Washington, DC 20560

NATIONAL MUSEUM OF AMERICAN HISTORY, SMITHSONIAN INSTITUTION
14th Street and Constitution Avenue NW
Washington, DC 20560

NATIONAL MUSEUM OF NATURAL HISTORY, SMITHSONIAN INSTITUTION
Department of Public Programs
10th Street and Constitution Avenue NW
Washington, DC 20560

Florida

BREVARD MUSEUM OF ART & SCIENCE, INC.
1463 Highland Avenue
Melbourne, FL 32935

CAPE CANAVERAL (p. 73)
(And while you're there, think of Orlando with its theme parks)

THE CHILDREN'S SCIENCE CENTER
2915 NE Pine Island Road
Cape Coral, FL 33909

CITY OF BOCA RATON CHILDREN'S SCIENCE EXPLORIUM
150 Crawford Boulevard
Boca Raton, FL 33432

DISCOVERY SCIENCE CENTER—CENTRAL FLORIDA COMMUNITY COLLEGE
50 South Magnolia Avenue
Ocala, FL 34474

EVERGLADES (p. 138)
Southern Florida

FAIRCHILD TROPICAL GARDEN
10901 Old Cutler Road
Miami, FL 33156

Florida Adventure Museum
260 West Retta Esplanade
Punta Gorda, FL 33950

Focus Center, Inc.
139 Brooks Street SE
Fort Walton Beach, FL 32548

Great Exploration, The Hands-on Museum
1120 4th Street South
St. Petersburg, FL 33701

GMIZ—Gulfcoast Wonder and Imagination Zone
8251 15th Street East
Sarasota, FL 34243

The Imaginarium Hands-on Museum and Aquarium
2000 Cranford Avenue
Fort Myers, FL 33916

Kennedy Space Center Visitor Complex
SR 405-Mail Code: DNPS
Kennedy Space Center, FL 32899

Lowry Park Zoo
7530 North Boulevard
Tampa, FL 33604

Miami Museum of Science & Space Transit Planetarium
3280 South Miami Avenue
Miami, FL 33129

Museum of Science & Industry
4801 East Fowler Avenue
Tampa, FL 33617

The Museum of Arts and Sciences
1040 Museum Boulevard
Daytona Beach, FL 32114

Museum of Discovery & Science, Inc.
401 SW Second Street
Fort Lauderdale, FL 33312

Museum of Science and History of Jacksonville, Inc.
1025 Museum Circle
Jacksonville, FL 32207

National Museum of Naval Aviation
1750 Radford Boulevard, Suite C
NAS Pensacola, FL 32508

Odyssey Science Center
345 South Magnolia Road, #B12
Tallahassee, FL 32301

Orlando Science Center
777 East Princeton Street
Orlando, FL 32803

South Florida Science Museum, Inc.
4801 Dreher Trail North
West Palm Beach, FL 33414

Withlacoochee Creek (p. 163)
West Florida Peninsula between Levy and Citrus counties, FL

Georgia

Fernbank Science Center
156 Heaton Park Drive NE
Atlanta, GA 30307

Georgia Southern University Museum
Southern Drive
Statesboro, GA 30460

The Museum of Arts & Sciences
4182 Forsyth Road
Macon, GA 31210

NATIONAL SCIENCE CENTER (FORT DISCOVERY)
One Seventh Street on Riverwalk
Augusta, GA 30901

SCITREK, THE SCIENCE AND TECHNOLOGY MUSEUM OF ATLANTA
395 Piedmont Avenue NE
Atlanta, GA 30308

Hawaii

BISHOP MUSEUM
1525 Bernice Street
Honolulu, HI 96817

KECK OBSERVATORY (p. 52)
Mauna Kea, HI

Idaho

THE DISCOVERY CENTER OF IDAHO, INC.
131 Myrtle Street
Boise, ID 83702

HAGEMAN FOSSIL BEDS NATIONAL MONUMENT (p. 162)
Hagerman, ID

IDAHO SCIENCE CENTER
P.O. Box 16
Arco, ID 83213

Illinois

THE ADLER PLANETARIUM AND ASTRONOMY MUSEUM
1300 South Lake Shore Drive
Chicago, IL 60605

ARGONNE NATIONAL LABORATORY (p. 19)
9700 S. Cass Avenue
Argonne, IL 60439

THE BURPEE MUSEUM OF NATURAL HISTORY
737 North Main Street
Rockford, IL 61103

CHICAGO ACADEMY OF SCIENCES, THE NATURE MUSEUM
2060 North Clark Street
Chicago, IL 60614

Chicago Botanic Garden
1000 Lake Cook Road
Glencoe, IL 60022

Chicago Children's Museum
Navy Pier, 700 East Grand Avenue, Suite 127
Chicago, IL 60611

Chicago River and Locks (p. 133)
Chicago, IL

Discovery Center Museum
711 North Main Street
Rockford, IL 61103

Fermi National Accelerator Laboratory—Leon M. Lederman Science, Education Center (p. 14)
Kirk and Pine Streets
Batavia, IL 60510

The Field Museum of Natural History (p. 152)
1200 South Lake Shore Drive
Chicago, IL 60605

Illinois State Museum
Spring and Edwards Streets
Springfield, IL 62706-5000

International Museum of Surgical Science
1524 North Lake Shore Drive
Chicago, IL 60610

JFK Health World
1301 South Grove Avenue
Barrington, IL 60010

Lakeview Museum of Arts and Sciences
1125 West Lake Avenue
Peoria, IL 61614

LINCOLN PARK ZOO
2001 N Clark Street
Chicago, IL 60614

MUSEUM OF SCIENCE AND INDUSTRY (p. 165)
57th Street and Lake Shore Drive
Chicago, IL 60637

THE SCIENCE CENTER
1237 E. Main Street, Space C-2
Carbondale, IL 62901

SCITECH, SCIENCE AND TECHNOLOGY INTERACTIVE CENTER
18 West Benton Street
Aurora, IL 60506

SHEDD AQUARIUM
South Lake Shore Drive
Chicago, IL.

Indiana

THE CHILDREN'S MUSEUM OF INDIANAPOLIS
3000 North Meridian Street
Indianapolis, IN 46208

CHILDREN'S SCIENCE & TECHNOLOGY MUSEUM OF TERRE HAUTE
523 Wabash Avenue
Terre Haute, IN 47807

EVANSVILLE MUSEUM OF ARTS AND SCIENCE
411 Southeast Riverside Drive
Evansville, IN 47713

IMAGINATION STATION
600 North 4th Street
Lafayette, IN 47901

INDIANAPOLIS ZOOLOGICAL SOCIETY, INC.
1200 West Washington Street
Indianapolis, IN 46222

Muncie Children's Museum
515 South High Street
Muncie, IN 47305

Science Central
1950 North Clinton Street
Fort Wayne, IN 46805

Iowa

Family Museum of Arts and Science
2900 Learning Campus Drive
Bettendorf, IA 52722

Flood Museum
Santa Fe R.R. Station
Fort Madison, IA

Grout Museums: Bluedorn Science Imaginarium
503 South Street
Waterloo, IA 50701

Iowa City Area Science Center, Inc.
Old Capital Mall, Clinton Street & Washington Street
Iowa City, IA 52245

Putnam Museum of History and Natural Science
1717 West 12th Street
Davenport, IA 52804

Science Center of Iowa
4500 Grand Avenue
Des Moines, IA 50312

Science Station
427 First Street SE
Cedar Rapids, IA 52401

University Museum
University of Northern Iowa
3219 Hudson Road
Cedar Falls, IA 50614

Kansas

CHILDREN'S MUSEUM OF WICHITA
435 South Water
Wichita, KS 67202

STERNBERG MUSEUM OF NATURAL HISTORY
Fort Hays State University
600 Park Street
Hays, KS 67601

UNIVERSITY OF KANSAS NATURAL HISTORY MUSEUM
Dyche Hall
The University of Kansas
Lawrence, KS 66045

Kentucky

KENTUCKY HIGHLANDS MUSEUM
1620 Winchester Avenue
Ashland, KY 41101

LOUISVILLE SCIENCE CENTER
727 West Main Street
Louisville, KY 40202

MAMMOTH CAVE (p. 150)
Cave City, KY

Louisiana

AUDUBON INSTITUTE (AUDUBON ZOOLOGICAL GARDEN AQUARIUM OF THE AMERICAS, ENTERGY IMAX THEATRE, AND *LOUISIANA NATURE CENTER)*
6500 Magazine Street
New Orleans, LA 70118

DELTA (p. 123)
Southern Louisiana
Mississippi River

Levee and the Aquarium (p. 124)
New Orleans, LA

Louisiana Arts and Science Center
100 South River Road
Baton Rouge, LA 70802

Louisiana Children's Museum
420 Julia Street
New Orleans, LA 70130

Louisiana State University Museum of Natural Science
119 Foster Hall
Baton Rouge, LA 70803

Sci-Port Discovery Center
820 Clyde Fant Parkway
Shreveport, LA 71101

Maine

The Children's Museum of Maine
142 Free Street
Portland, ME 04101

Maryland

Maryland Science Center
601 Light Street
Baltimore, MD 21230

National Aquarium in Baltimore
501 East Pratt Street, Pier 3
Baltimore, MD 21202

Massachusetts

The Children's Museum
Museum Wharf
300 Congress Street
Boston, MA 02210

The Computer Museum
300 Congress Street
Boston, MA 02210

EcoTarium
222 Harrington Way
Worcester, MA 01604

Harvard-Smithsonian Observatory (p. 46)
Cambridge, MA

Museum of Science
Science Park
Boston, MA 02114

National Plastics Center and Museum
210 Lancaster Street
Leominster, MA 01453

New england Aquarium
Central Wharf
Boston, MA 02110

Springfield Science Museum
At the Quadrangle, Corner of State & Chestnut streets
Springfield, MA 01103

Woods Hole Oceanographic Institution exhibit Center
Mail Stop #5
Woods Hole, MA 02543

Michigan

Alfred P. Sloan Museum
1221 East Kearsley Street
Flint, MI 48503

Ann Arbor Hands-On Museum
219 East Huron Street
Ann Arbor, MI 48104

Cranbrook Institute of Science
1221 North Woodward Avenue
Bloomfield Hills, MI 48303

DETROIT SCIENCE CENTER
5020 John R. Street
Detroit, MI 48202

EXHIBIT MUSEUM OF NATURAL HISTORY
1109 Geddes Avenue
Ann Arbor, MI 48109

THE FLINT CHILDREN'S MUSEUM
1602 West Third Avenue
Flint, MI 48504

HENRY FORD MUSEUM & GREENFIELD VILLAGE
20900 Oakwood Boulevard
Dearborn, MI 48124

HALL OF IDEAS, MIDLAND CENTER FOR THE ARTS, INC.
1801 West St. Andrews Road
Midland, MI 48640

IMPRESSION 5 SCIENCE CENTER
200 Museum Drive
Lansing, MI 48933

KALAMAZOO VALLEY MUSEUM
230 North Rose Street
Kalamazoo, MI 49007

KINGMAN MUSEUM OF NATURAL HISTORY
West Michigan Avenue at 20th Street
Battle Creek, MI 49017

MICHIGAN SPACE AND SCIENCE CENTER
2111 Emmons Road
Jackson, MI 49201

SOUTHWESTERN MICHIGAN COLLEGE MUSEUM
58900 Cherry Grove Road
Dowagiac, MI 49047

Minnesota

The Bakken Library and Museum
3537 Zenith Avenue South
Minneapolis, MN 55416

Headwaters Science Center
413 Beltrami Avenue
Bemidji, MN 56601

Heritage Hjemkomst Interpretive Center
202 1st Avenue North
Moorhead, MN 56560

Minnesota Children's Museum
10 West 7th Street
St. Paul, MN 55102

Science Museum of Minnesota
30 East 10th Street
St. Paul, MN 55101

Mississippi

Davis Planetarium
201 East Pascagoula
Jackson, MS 39205

Mississippi Museum of Natural Science
111 North Jefferson Street
Jackson, MS 39202

Missouri

Discovery Center of Springfield, Inc.
438 East St. Louis Street
Springfield, MO 65806

Kansas City Museum/Science City at Union Station
3218 Gladstone Boulevard
Kansas City, MO 64123

THE MAGIC HOUSE, ST. LOUIS CHILDREN'S MUSEUM
516 South Kirkwood Road
St. Louis, MO 63110

MISSOURI BOTANICAL GARDEN
4344 Shaw Avenue
St. Louis, MO 63166

TOM SAWYER'S CAVE (p. 150)
Hannibal, MO

ST. LOUIS SCIENCE CENTER
5050 Oakland Avenue
St. Louis, MO 63110

ST. LOUIS ZOO
Forest Park
St. Louis, MO 63110

SUBTROPOLIS (p. 152)
Kansas City, MO

Montana

MUSEUM OF THE ROCKIES
Montana State University
600 West Kagy Boulevard
Bozeman, MT 59717

Nebraska

EDGERTON EXPLORIT CENTER
208 16th Street
Aurora, NE 68818

OMAHA CHILDREN'S MUSEUM
500 South 20th Street
Omaha, NE 68102

UNIVERSITY OF NEBRASKA STATE MUSEUM
14th and U Streets
Lincoln, NE 68588

Nevada

LIED DISCOVERY CHILDREN'S MUSEUM
833 Las Vegas Boulevard North
Las Vegas, NV 89101

WILBUR MAY CENTER —WASHOE COUNTY PARKS AND RECREATION DEPARTMENT
1502 Washington Street
Reno, NV 89503

New Hampshire

CHRISTA MCAULIFFE PLANETARIUM
3 Institute Drive
Concord, NH 03301

SEE SCIENCE CENTER
324 Commercial Street
Manchester, NH 03101

New Jersey

EINSTEIN MUSEUM (p. 24)
Princeton, NJ

INVENTION FACTORY SCIENCE CENTER
650 South Broad Street
Trenton, NJ 08611

LIBERTY SCIENCE CENTER
251 Phillip Street
Jersey City, NJ 07305

THE NEWARK MUSEUM AND DREYFUS PLANETARIUM
49 Washington Street
P.O. Box 540
Newark, NJ 07101

THOMAS H. KEAN NEW JERSEY STATE AQUARIUM AT CAMDEN
1 Riverside Drive
Camden, NJ 08103

New Mexico

Blackwater Draw (p. 157)
Clovis, NM

Bradbury Science Museum, Los Alamos National Laboratory
15th and Central
Los Alamos, NM 87545

Carlsbad Cavern (p. 151)
Alburquerque area, NM

Explora Science Center and Children's Museum
800 Rio Grande NW, Suite 19
Albuquerque, NM 87104

Las Cruces Museum of Natural History
700 South Telshor Boulevard
Mesilla Valley Mall
Las Cruces, NM 88011

LodeStar
801 University Boulevard SE, Suite 301
Albuquerque, NM 87106

New Mexico Museum of Natural History and Science
1801 Mountain Road NW
Albuquerque, NM 87104

Sandia National Laboratory (p. 20)
Albuquerque area, NM

Space Center
Top of New Mexico Highway 2001
Alamogordo, NM 88310

New York

American Museum of Natural History (p. 166)
Central Park West at 79th Street
New York, NY 10024

BNL Science Museum
Brookhaven National Laboratory
Upton, NY 11973

Brookhaven National Laboratory (p. 18)
New York City, NY

Brooklyn Botanic Garden
1000 Washington Avenue
Brooklyn, NY 11225

The Brooklyn Children's Museum
145 Brooklyn Avenue
Brooklyn, NY 11213

Buffalo Museum of Science
1020 Humboldt Parkway
Buffalo, NY 14211

Central Park Rocks (p. 142)
New York City, NY

Consolidated Edison Underground Room (p. 154)
14th Street on Third Avenue
New York City, NY 10003

DNA Learning Center
334 Main Street
Cold Spring Harbor, NY 11724

Erie Canal (p.132)
upstate New York

Rose Center for Earth and Space
81st Street (at Central Park West)
New York City, NY 10024

Hudson River Museum
511 Warburton Avenue
Yonkers, NY 10701

LIMSAT (Long Island Museum of Science and Technology)
P.O. Box 804
Melville, NY 11747

MID-HUDSON CHILDREN'S MUSEUM
South Hills Mall
838 South Road, Suite 155
Poughkeepsie, NY 12601

MILTON J. RUBENSTEIN MUSEUM OF SCIENCE & TECHNOLOGY
500 South Franklin Street
Syracuse, NY 13202

NEW YORK AQUARIUM
Surf and W. 8th Streets
Brooklyn, NY 11224

NEW YORK BOTANICAL GARDEN
200th Street and Southern Boulevard
Bronx, NY 10458

NEW YORK HALL OF SCIENCE
47-01 111th Street, Flushing Meadows
Corona Park, NY 11368

NEW YORK STATE MUSEUM
Cultural Education Center, Room 3099
Albany, NY 12230

NEW YORK TRANSIT MUSEUM
Boerum Place and Schermerhorn Street
Brooklyn, NY 11201

NIAGARA FALLS (P. 140)
Niagara Falls, NY

PALEONTOLOGICAL RESEARCH INSTITUTION
1259 Trumansburg Road
Ithaca, NY 14850

ROBERSON MUSEUM AND SCIENCE CENTER
30 Front Street
Binghamton, NY 13905

ROCHESTER MUSEUM & SCIENCE CENTER
657 East Avenue
Rochester, NY 14607

SCHENECTADY MUSEUM
Nott Terrace Heights
Schenectady, NY 12308

SCIENCE AND DISCOVERY CENTER
Arnot Mall
3300 Chambers Road
Horseheads, NY 14844

SCIENCE DISCOVERY CENTER OF ONEONTA
Physical Science Building, State College
Oneonta, NY 13820

SCIENCE MUSEUM
250 Route 25A
Shoreham, NY 11786

SCIENCENTER
601 First Street
Ithaca, NY 14850

STATEN ISLAND CHILDREN'S MUSEUM
1000 Richmond Terrace
Staten Island, NY 10301

WILDLIFE CONSERVATION SOCIETY—BRONX ZOO/WILDLIFE CONSERVATION PARK
185th Street and Southern Boulevard
Bronx, NY 10460

North Carolina

THE ARTS & SCIENCE CENTER
1335 Museum Road
Statesville, NC 28677

CAPE FEAR MUSEUM
814 Market Street
Wilmington, NC 28401

CATAWBA SCIENCE CENTER
243 3rd Avenue NE
Hickory, NC 28603

Children's Museum of Iredell County
P.O. Box 223
134 Court Street
Statesville, NC 28687

Discovery Place, Inc.
301 North Tryon Street
Charlotte, NC 28202

The Health Adventure
2 South Pack Square
Asheville, NC 28801

Natural Science Center of Greensboro, Inc.
4301 Lawndale Drive
Greensboro, NC 27455

North Carolina Museum of Life and Science
433 Murray Avenue
Durham, NC 27704

North Carolina State Museum of Natural Sciences
102 North Salisbury Street
Raleigh, NC 27603

The Rocky Mount Children's Museum
1610 Gay Street
Rocky Mount, NC 27804

Schiele Museum of Natural History and Planetarium, Inc.
1500 East Garrison Boulevard
Gastonia, NC 28054

Science Museums of Wilson, Inc.
224 East Nash Street
Wilson, NC 27893

SciWorks, The Science Center and Environmental Park of Forsyth County
400 West Hanes Mill Road
Winston-Salem, NC 27105

North Dakota

Dakota Science Center
4300 Dartmouth Drive, Suite 185
Grand Forks, ND 58202

Gateway to Science Center
Gateway Mall, 2700 State Street
Bismarck, ND 58501

Ohio

Cincinnati Museum Center
1301 Western Avenue
Cincinnati, OH 45203

COSI, Ohio's Center of Science & Industry
280 East Broad Street
Columbus, OH 43215

COSI Toledo
1 Discovery Way
Toledo, OH 43604

The Dayton Museum of Discovery
2600 DeWeese Parkway
Dayton, OH 45414

The Great Lakes Science Center
601 Erieside Avenue
Cleveland, OH 44114

The Health Museum of Cleveland
8911 Euclid Avenue
Cleveland, OH 44106

Inventure Place, Home of the National Inventors Hall of Fame
221 South Broadway Street
Akron, OH 44308

Mc Kinley Museum of History, Science and Industry
800 McKinley Monument Drive NW
Canton, OH 44708

Rainbow Children's Museum & TRW Early Learning Center
10730 Euclid Avenue
Cleveland, OH 44106

SciMaTec—The University of Toledo
2801 West Bancroft Street
Toledo, OH 43623

Oklahoma

Harmon Science Center
5707 East 41st Street
Tulsa, OK 74135

Kirkpatrick Science and Air Space Museum at Omniplex
2100 NE 52nd Street
Oklahoma City, OK 73111

Oregon

A. C. Gilbert's Discovery Village
116 Marion Street NE
Salem, OR 97301

The High Desert Museum
59800 South Highway 97
Bend, OR 97702

Oregon Museum of Science and Industry
1945 SE Water Avenue
Portland, OR 97214

WISTEC, Willamette Science & Technology Center
2300 Leo Harris Parkway
Eugene, OR 97401

Pennsylvania

The Academy of Natural Sciences
1900 Benjamin Franklin Parkway
Philadelphia, PA 19103

The Carnegie Science Center
One Allegheny Avenue
Pittsburgh, PA 15212

The Discovery Center of Science & Technology
511 East Third Street
Bethlehem, PA 18015

The Franklin Institute
20th Street and the Parkway
Philadelphia, PA 19103

Hawk Mountain Sanctuary Association
1700 Hawk Mountain Road
Kempton, PA 19529

The Museum of Scientific Discovery
Strawberry Square
Third and Walnut streets
Harrisburg, PA 17108

North Museum of Natural History and Science
400 College Avenue
Lancaster, PA 17603

The Pittsburgh Children's Museum
10 Children's Way
Pittsburgh, PA 15212

Planetarium
Philadelphia, PA

Please Touch Museum
210 North 21st Street
Philadelphia, PA 19103

Reading Public Museum
500 Museum Road
Reading, PA 19611

The Whitaker Center for Science and the Arts
301 Market Street, 7th Floor
Harrisburg, PA 17101

Zoological Society of Philadelphia
3400 West Girard Avenue
Philadelphia, PA 19104

Puerto Rico

Arecibo Radio Observation (p. 61)
Puerto Rico

Rhode Island

Roger Williams Park Zoo
1000 Elmwood Avenue
Providence, RI 02905

Thames Science Center
77 Long Wharf
Newport, RI 02804

South Carolina

Roper Mountain Science Center
504 Roper Mountain Road
Greenville, SC 29615

South Carolina State Museum
301 Gervais Street
Columbia, SC 29201

South Dakota

National Motorcycle Museum & Hall of Fame
2438 South Junction Avenue
Sturgis, SD 57785

South Dakota Discovery Center & Aquarium
805 West Sioux Avenue
Pierre, SD 57501

Tennessee

American Museum of Science & Energy
300 South Tulane Avenue
Oak Ridge, TN 37830

THE CHILDREN'S MUSEUM OF MEMPHIS
2525 Central Avenue
Memphis, TN 38104

THE CREATIVE DISCOVERY MUSEUM
321 Chestnut Street
Chattanooga, TN 37402

CUMBERLAND SCIENCE MUSEUM
800 Fort Negley Boulevard
Nashville, TN 37203

EAST TENNESSEE DISCOVERY CENTER
516 North Beaman Street
Chilhowee Park
Knoxville, TN 37914

HANDS ON! REGIONAL MUSEUM
315 East Main Street
Johnson City, TN 37601

HANDS-ON SCIENCE CENTER
101 Mitchell Boulevard
Tullahoma, TN 37388

MEMPHIS MUSEUM SYSTEM
3050 Central Avenue
Memphis, TN 38111

MODEL PARTHENON
Nashville, TN

TENNESSEE AQUARIUM
One Broad Street
P.O. Box 11048
Chattanooga, TN 37401

Texas

AUSTIN CHILDREN'S MUSEUM
201 Colorado Street
Austin, TX 78701

Austin Nature Center & Science Center
301 Nature Center Drive
Austin, TX 78746

Children's Museum of Houston
1500 Binz
Houston, TX 77004

The Cook Center
3100 West Collin Avenue
Corsicana, TX 75110

Dallas Museum of Natural History
3535 Grand Avenue
Dallas, TX 75226

Davis Mountain Observatory (p. 59)
Davis Mountain, TX

The Discovery Science Place
308 North Broadway
Tyler, TX 75702

Fort Worth Museum of Science and History
1501 Montgomery Street
Fort Worth, TX 76107

Don Harrington Discovery Center
1200 Streit Drive
Amarillo, TX 79106

Houston Museum of Natural Science
One Hermann Circle Drive
Houston, TX 77030

Johnson Space Center (p. 76)
210 NASA Rd. 1
Houston, TX 77058

McAllen International Museum
1900 Nolana
McAllen, TX 78504

McDonald Observatory Visitor's Center
Highway 118 North
Fort Davis, TX 79734

The Museum of Health and Medical Science
1515 Hermann Drive
Houston, TX 77004

Science Spectrum
2579 South Loop 289, Suite 250
Lubbock, TX 79423

Southwest Museum of Science & Technology
1318 Second Avenue
Dallas, TX 75210

Texas State Aquarium
2710 North Shoreline Boulevard
Corpus Christi, TX 78402

Witte Museum
3801 Broadway
San Antonio, TX 78209

Utah

Dinosaur National Monument
Vernal, UT

Utah Museum Natural History
University of Utah
1390 East President's Circle
Salt Lake City, UT 84112

Vermont

Fairbanks Museum and Planetarium
Main and Prospect streets
St Johnsbury, VT 05819

LAKE CHAMPLAIN BASIN SCIENCE CENTER
One College Street
Burlington, VT 05401

MONTSHIRE MUSEUM OF SCIENCE
One Montshire Road
Norwich, VT 05055

Virginia

CHALLENGER CENTER FOR SPACE SCIENCE EDUCATION
1029 North Royal Street, Suite 300
Alexandria, VA 22314

LURAY, SHENANDOAH, SKYLINE, NATURAL BRIDGE
CAVERNS

MATHEMATICS & SCIENCE CENTER
2401 Harman Street
Richmond, VA 23223

NAUTICUS, THE NATIONAL MARITIME CENTER
One Waterside Drive
Norfolk, VA 23510

SCIENCE MUSEUM OF VIRGINIA
2500 West Broad Street
Richmond, VA 23220

SCIENCE MUSEUM OF WESTERN VIRGINIA
One Market Square
Roanoke, VA 24011

SHENANDOAH VALLEY DISCOVERY MUSEUM
54 South Loudoun Street
Winchester, VA 22601

VIRGINIA AIR AND SPACE CENTER AND HAMPTON ROADS HISTORY CENTER
600 Settlers Landing Road
Hampton, VA 23669

VIRGINIA DISCOVERY MUSEUM
524 East Main
Charlottesville, VA 22902

VIRGINIA LIVING MUSEUM
524 J. Clyde Morris Boulevard
Newport News, VA 23601

VIRGINIA MUSEUM OF NATURAL HISTORY
1001 Douglas Avenue
Martinsville, VA 24112

Washington

COLUMBIA RIVER EXHIBITION OF HISTORY, SCIENCE AND TECHNOLOGY (CREHST)
825 Jadwin Avenue, A1-60
Richland, WA 99352

THE FUNHOUSE
Route 1, Box 626
Eastsound, WA 98245

MUSEUM OF FLIGHT
9404 East Marginal Way South
Seattle, WA 98108

MOUNT RAINIER

MOUNT ST. HELENS (p. 102)

PACIFIC SCIENCE CENTER
200 Second Avenue North
Seattle, WA 98109

STONEROSE INTERPRETIVE CENTER
61 North Kean Street
Republic, WA 99166

West Virginia

SCIENCE CENTER OF WEST VIRGINIA
500 Bland Street
Bluefield, WV 24701

SUNRISE MUSEUM
746 Myrtle Road
Charleston, WV 25314

Wisconsin

BIOTREK: THE MUSEUM OF LIVING SCIENCES
425 Henry Mall
Madison, WI 53706

DISCOVERY WORLD: THE JAMES LOVELL MUSEUM OF SCIENCE, ECONOMICS, & TECHNOLOGY
712 West Wells Street
Milwaukee, WI 53233

THE IMAGINARIUM OF RACINE
500 Monument Square
Racine, WI 53403

MILWAUKEE PUBLIC MUSEUM
800 West Wells Street
Milwaukee, WI 53208

YERKES OBSERVATORY (p. 50)
Williams Bay, WI

Wyoming

THE WYOMING ADVENTURE CENTER
East Collins
Casper, WY 82601

CANADA

BAY OF FUNDY

BIÔDOME DE MONTRÉAL
4777 Avenue Pierre-de-Coubertin
Montréal, Québec H1V 1B3

CALGARY SCIENCE CENTRE
701–11 Street SW
Calgary, Alberta T2P 2C4

CANADIAN MUSEUM OF NATURE
240 McLeod Street
Victoria Memorial Museum Building
Ottawa, Ontario K1P 6P4

DISCOVERY CENTRE
1593 Barrington Street
Halifax, Nova Scotia B3J 1Z7

EDMONTON SPACE & SCIENCE CENTRE
11211 142nd Street NW
Edmonton, Alberta T5M 4A1

NATIONAL MUSEUM OF SCIENCE & TECHNOLOGY CORPORATION
1867 St. Laurent Boulevard
Ottawa, Ontario K1G 5A3

NEUTRINO TELESCOPE (p. 70)
Sudbury, Ontario

THE OLD PORT OF MONTREAL
333 Rue de la Commune Ouest
Montréal, Québec H2Y 2E2

ONTARIO SCIENCE CENTRE
770 Don Mills Road
North York, Ontario M3C 1T3

ONTARIO SCIENCE MUSEUM
Toronto, Ontario

ROYAL TYRRELL MUSEUM
Drumhoolie, Alberta

SASKATCHEWAN SCIENCE CENTRE, INC.
In Wascana Centre
2903 Powerhouse Drive
Regina, Saskatchewan S4P 3M3

Science North
100 Ramsey Lake Road
Sudbury, Ontario P3E 5S9

Science World British Columbia
1455 Quebec Street
Vancouver, British Columbia V6A 3Z7

MEXICO

Centro de Ciencias de Sinaloa
Avenida de las Americas 2771 Norte
Culiacán, Sinaloa 80010

Centro de Ciencias 'EXPLORA'
Boulevard Francisco Villa 202
Leo, Guanajuato 37500

Descubre, Interactive Museum of Science and Technology
Avenida San Miguel
S.N. Aguascalientes
Ags C.P. 20270

Museo Tecnológio de la Comisón Federal de Electricidad
2a. Seccion
Bosque Chapultepec
Mexico City, D.F. C.P. 11870

Papalote, Museo del Niño
Avenida Constituyentes 268
Col. Daniel Garza
Mexico City D.F. 1111

Universum, Museo de la Ciencias, UNAM
Insurgentes Sur 3000, Centro C
Cuidad Universitaria-Curcuito Cultural
Mexico City, D.F. 04510

ARGENTINA

EUREKA CENTRO INTERACTIVO CIENTIFICO
Uriburu S/N Parque Gral. San
Martin
Mendoza 5500

MUSEO PARTICIPATIVO DE CIENCIAS
Junin 1930
Buenos Aires, BS AS 1113

AUSTRALIA

AYERS ROCK

DARWIN MUSEUM
Darwin

GREAT BARRIER REEF

THE INVESTIGATOR SCIENCE & TECHNOLOGY CENTRE
Rose Terrace
Wayville, South Australia 5034

POWERHOUSE MUSEUM OF APPLIED ARTS & SCIENCES
500 Harris Street
Ultimo, N.S.W. 2007

QUEENSLAND SCIENCENTRE
110 George Street
Brisbane, Queensland 4000

QUESTACON, THE NATIONAL SCIENCE AND TECHNOLOGY CENTRE
King Edward Terrace
Parkes, ACT 2600

RADIO OBSERVATORY
Triabunna, Tasmania

SCITECH DISCOVERY CENTRE
Corner Railway Parade and Sutherland Street
1st Floor, City West
West Perth, Western Australia 6005

AUSTRIA

TECHNISCHES MUSEUM WIEN (TECHNOLOGY MUSEUM OF VIENNA)
Mariahilferstrasse 212
A-1140 Wien

BELGIUM

F.T.I. FOUNDATION (STICHING FLANDERS TECHNOLOGY INTERNATIONAL)
Stormstraat 1
Brussels 1000

BERMUDA

BERMUDA UNDERWATER EXPLORATION INSTITUTE
40 Crow Lane

BRAZIL

CATARACTS OF IGUAÇÓ (p. 140)

FUNDAÇÃO, PLANETÁRIO DO RIO DE JANEIRO
Avenida Padre Leonel Franca 240
Gavea
Rio de Janeiro, RJ 22451-000

CHILE

Fundacion Tiempos Nuevos
Hindenburg 709
Santiago

La Silla Observatories (p. 59)

CHINA

Great Wall
Beijing

Hong Kong Science Museum
2 Science Museum Road
Tsim Sha Tsui East
Kowloon, Hong Kong

COLOMBIA

Maloka-Centro Interactivo de Ciencia y Tecnología
Carrera 50 No. 27–70
Edificio Camilo Torres
Modulo 3
Santa Fe de Bogota D.C.

COSTA RICA

Fundacion para el Centro Nacional de Ciencia y Tecnología
Calles 6 y 8, Avenida Segunda
San Jose

DENMARK

Experimentarium
Tuborg Havnevej 7
DK-2900 Hellerup

FINLAND

Heureka, The Finish Science Centre
Tiedepuisto 1
Vantaa FIN-01300

FRANCE

Caves of Lascaux (p. 149)
Lascaux, Dordogne

Cité des Sciences et de l'Industrie
30 Avenue Corentin-Cariou
Paris 75019

Fondation Enterprise Réussite Scolaire
4 Rue Joseph Serlin
Lyon 69001

Palais de la Decouverte
Avenue Franklin D. Roosevelt
Paris 75008

GERMANY

Deutsches Hygiene-Museum
Lingnerplatz 1
D-01069 Dresden

Deutsches Museum (p. 52)
Museumsinsel 1
Munchen, CO 80538

GREECE

Delphic Oracle Cave (p. 148)
Mount Parnasus, Athens

Technology Museum of Thessaloniki
Building 47, 2nd Street
Industrial Area of Thessaloniki Sindos
Thessaloniki 57022

ICELAND

Geysers (p. 108)

Tectonic Plane Boundary (p. 109)

INDIA

Birla Industrial and Technological Museum
19A Gurusaday Road
Calcutta 700 019

National Science Centre
Pragati Maidan, Near Gate #1
Bhairon Road
New Delhi 110 001

Nehru Science Center
Dr. E. Moses Road
Worli
Mumbai, Maharashtra 400 018

Visvesvaraya Industrial & Technological Museum
Kasturba Road
Bangalore 560 001

INDONESIA

Sasana Budaya Ganesa Institut Teknologi Bandung
Jalan Ganesa No. 10
Bandung 40132

Science and Technology Center of Indonesia
Taman Mini Indonesia Indah
Jakarta 13560

IRAN

Zirakzadeh Science Foundation
112 Vanak Avenue
Tehran 19919

ISRAEL

Bloomfield Science Museum, Jerusalem
Museum Boulevard
Jerusalem 91904

Garden of Science
Weizmann Institute of Science
Youth Activities Section
P.O. Box 26
Rehovot 76100

Israel National Museum of Science, Planning & Technology/Recanati Center
Shmaryahu Levin Street
Haifa 31448

ITALY

Aqueducts (p. 130)
Rome

MUSEO ARCHEOLOGICO REGINALE "PAOLO ORSI" (p. 171)
Piazza Promo 14
Syracuse, Sicily

MOUNT ETNA (p. 103)
Sicily

MOUNT VESUVIUS (p. 100)

POMPEII (p. 172)

JAMAICA

ICWI GROUP FOUNDATION
2 Saint Lucia Avenu
Kingston 5

JAPAN

HIROSHEMA PEACE PARK
Hiroshima

NAGOYA CITY SCIENCE MUSEUM
17-1 Sakae 2-Chome, Naka-ki
Nagoya 460

NATIONAL SCIENCE MUSEUM
Ueno Koen 7-20, Taito-ku
Tokyo 110-8718

SCIENCE MUSEUM OF THE JAPAN SCIENCE FOUNDATION
2-1 Kitanomaru-Koen Chiyoda-ku
Tokyo 102

TSUKUBA EXPO CENTER
2-9 Azuma
Tsukuba Ibaraki Prefect 305-0031

KENYA

Animal Reserves
Masai Mara, etc.

National Museum of Kenya (p. 166)
P.O. Box 40658
Nairobi

Natural History Museum (p. 166)
Nairobi

Railroad Museum (p. 166)
Nairobi

KUWAIT

International Discovery Co.
P.O. Box 27068 SAFAT
Kuwait 13131

Kuwait Science Club
P.O. Box 23259
Safat 13093

MALAYSIA

National Science Centre of Malaysia
Jalan Persiaran Bukit Kiara
50662 Kuala Lumpur

Petrosains & Concert Hall, PMT
Lot 422-425
Suria KLCC
Kuala Lumpur City Center,
50088 Kuala Lumpur

THE NETHERLANDS

New Metropolis Science and Technology Center
Oosterdok 2
1011 VX Amsterdam

NEW ZEALAND

Discovery World
419 Great King Street
Dunedin

Science Alive!
Moorhouse Avenue
Christchurch

Science Centre & Manawatu Museum
396 Main Street
Palmerston North

NORWAY

Teknoteket, National Science Center
Konvallvn 2B
0855 Oslo

PAKISTAN

National Museum of Science and Technology
Near Engineering University
G.T. Road
Lahore 54890

PANAMA

Canal (p. 134)

Science & Art Center
Despacho de la Primera Dama
Zona 1

PERU

Ruins (p. 173)

Machu Picchu

Cuzco

PHILIPPINES

Bicol Science Centrum
J. Miranda Avenue
Naga City 4400

Philippine Science Centrum
UP-Manila Compound
Pedro Gil Street, Ermita
Manila 1000

Quest, Center for Earth Science and Discovery
Goldencrest Building, Ground Floor
Ayala Center
1226 Makati City

RUSSIA

Space Museum (p. 84)
Moscow

STARLING MUSEUM (p. 83)
Star City

SAUDI ARABIA

JEDDAH SCIENCE AND TECHNOLOGY CENTRE
Corniche Street
Jeddah 21461

SCIENCE OASIS
P.O. Box 94501
Riyadh 11614

SINGAPORE

SINGAPORE SCIENCE CENTRE
Science Centre Road
Singapore 609081

SOUTH KOREA

NATIONAL SCIENCE MUSEUM OF KOREA
32-2 Kusung-Dong, Yusung-ku
Tae-Jeon 305-338

SPAIN

CASA DE LAS CIENCIAS AND DOMUS
Parque de Santa Margarita, S/N
La Coruña 15005

THE CITY OF ARTS AND SCIENCE
Arzobispo Mayoral 14
Valencia 46002

MUSEU DE LA CIENCIA DE LA FUNDACIO "LA CAIXA"
Teodor Roviralta 55
Barcelona 08022

SWEDEN

HIMMEL OCH HAV SCIENCE CENTER
Linkopings Universitet
Holmentorget 10
601.74 Norrkoping

TEKNIKENS HUS (HOUSE OF TECHNOLOGY)
Hogskoleomradet University
Lulea S-971 87

TEKNISKA MUSEET—TEKNORAMA NATIONAL MUSEUM OF SCIENCE AND TECHNOLOGY
Museivägen 7
Norra Djurgården
Stockholm S-115 93

SWITZERLAND

WATERFALLS (p. 140)
between Zurich and Italian border

TAIWAN

NATIONAL MUSEUM OF NATURAL SCIENCE
1 Kuan Chien Road
Taichung

NATIONAL SCIENCE & TECHNOLOGY MUSEUM
4F, 95 Pyng-Teng Road 80733
Kaohsiung

NATIONAL TAIWAN SCIENCE EDUCATION CENTRE
41 Nan-Hai Road
Taipei 100

TRINIDAD

National Science Centre (NIHERST)
Cor. Piarco Old Road and Churchill Roosevelt Highway
Piarco

TURKEY

Great Cistern
Istanbul

Science Center Foundation
Barbados Bulvari, Hasfirin Cad.
67 Sinanpasa Is Merkezi Kat 5
Besikas
806090 Istanbul

Tübitak, The Scientific and Technical Research Council of Turkey
Ataturk Bulvari #221
Kavaklidere
Ankara 06100

UNITED KINGDOM

British Museum
London
England

British War Museum
London
England

Cutty Sark
Greenwich
England

Eureka! The Museum for Children
Discovery Road
Halifax HX1 2NE
England

The Exploratory
Temple Meads
Bristol BS1 6QU
England

Greenwich Observatory
Greenwich
England

Hadrian's Wall (p. 168)
near the Scottish border
England

The International Centre for Life
Market Keepers House
Times Square, Scotswood Road
Newcastle upon Tyne NE1 4EP
England

Museum of the City of London
London
England

Natural History Museum (p. 24)
London
England

National Museums & Galleries on Merseyside
127 Dale Street
Liverpool L69 3LA
England

Royal Armouries
Armouries Drive
Leeds, Yorkshire LS10 1LT
England

SCIENCE MUSEUM, NATIONAL MUSEUM OF SCIENCE & INDUSTRY
Exhibition Road South
Kensington
London SW7 2DD
England

SCIENCE PROJECTS
20 St. James Street
Hammersmith
London W6 9RW
England

SOUTHAMPTON ROW
London
England

STONEHENGE (p. 167)
near Salisbury
England

TECHNIQUEST
Stuart Street
Cardiff CF1 6BW
Wales

THAMES BARRIER
Greenwich
England

ULSTER MUSEUM
Botanic Gardens
Belfast BT9 5AB
Northern Ireland

VENICE CANAL (p. 132)
London
England

VISITOR CENTRE, ROYAL OBSERVATORY
Blackford Hill
Edinburgh EH9 3HJ
Scotland

THE WELLCOME CENTRE FOR MEDICAL SCIENCE
210 Euston Road
London NW1 2BE
England

WOOLWICH ARSENAL
Greenwich
England

URUGUAY

ESPACIO CIENCIA
Avenida Italia 6201
Montevideo 11500

EXPLORA URUGUAY
Rambla Peru 1435
Montevideo 11300

VENEZUELA

MUSEO DE LOS NIÑOS DE CARACAS
Parque Central
Nivel Bolívar
Caracas 1010

Index

ABOUT THE AUTHOR

A multiple Hugo and Nebula Award–winning author, FREDERIK POHL has done just about everything one can do in the science fiction field. His most famous work is undoubtedly the novel *Gateway*, which won the Hugo, Nebula, and John W. Campbell Memorial awards for Best SF novel. *Man Plus* won the Nebula Award. His mature work is marked by a serious intellectual agenda and strongly held sociopolitical beliefs, without sacrificing narrative drive. In addition to his successful solo fiction, Pohl has collaborated with a variety of writers, including C. M. Kornbluth and Jack Williamson. The Pohl/Kornbluth collaboration, *The Space Merchants*, is a longtime classic of satiric science fiction. *The Starchild Trilogy* with Williamson is one of the more notable collaborations in the field. Pohl became a magazine editor when he was still a teenager. In the 1960s he piloted *Worlds of If* to three successive Hugos for Best Magazine. He also has edited original-story anthologies, including the early and notable *Star* series of the early 1950s. He has at various times been a literary agent, an editor of lines of science fiction books, and president of the Science Fiction Writers of America. For a number of years he has been active in the World SF movement. He and his wife, Elizabeth Anne Hull, a prominent academic active in the Science Fiction Research Association, live outside Chicago, Illinois.